PRACTICAL BUILDING CONSERVATION
VOLUME 4
METALS

Practical Building Conservation Series:

PRACTICAL BUILDING CONSERVATION

English Heritage Technical Handbook

VOLUME 4

METALS

John Ashurst
Nicola Ashurst
with
Geoff Wallis
and
Dennis Toner

Photographs by Nicola Ashurst
Graphics by Iain McCaig

Gower Technical Press

Published by
Gower Technical Press Ltd,
Gower House,
Croft Road,
Aldershot,
Hants GU11 3HR,
England

Reprinted 1989, 1991, 1993

British Library Cataloguing in Publication Data

Ashurst, John,
    Practical building conservation:
    English Heritage technical handbook.
    Vol. 4: Metals
    1. Great Britain. Buildings of historical
    importance. Conservation. Techniques
    I. Title   II. Ashurst, Nicola
    720'.28'8

ISBN 0 291 39748 4

Printed and bound in Great Britain by
BPC Wheatons Ltd, Exeter

CONTENTS

FOREWORD

by Peter Rumble CB, Chief Executive, English Heritage

Over many years the staff of the Research, Technical and Advisory Service of English Heritage have built up expertise in the theory and practice of conserving buildings and the materials used in buildings. Their knowledge and advice have been given mainly in respect of individual buildings or particular materials. The time has come to bring that advice together in order to make available practical information on the essential business of conserving buildings — and doing so properly. The advice relates to most materials and techniques used in traditional building construction as well as methods of repairing, preserving and maintaining our historic buildings with a minimum loss of original fabric.

Although the five volumes which are being published are not intended as specifications for remedial work, we hope that they will be used widely by those who write, read or use such specifications. We expect to revise and enlarge upon some of the information in subsequent editions as well as introducing new subjects. Although our concern is with the past, we are keenly aware that building conservation is a modern and advancing science to which we intend, with our colleagues at home and abroad, to continue to contribute.

The Practical Building Conservation Series

The contents of the five volumes reflect the principal requests for information which are made to the Research, Technical and Advisory Services of English Heritage (RTAS) in London.

RTAS does not work in isolation; it has regular contact with colleagues in Europe, the Americas and Australia, primarily through ICOMOS, ICCROM and APT. Much of the information is of direct interest to building conservation practitioners in these continents as well as their British counterparts.

English Heritage

English Heritage, the Historic Buildings and Monuments Commission for England came into existence on 1st April 1984, set up by the Government but independent of it. Its duties cover the whole of England and relate to ancient monuments, historic buildings, conservation areas, historic gardens and archaeology. The Commission consists of a chairman and up to sixteen other members. Commissioners are appointed by the Secretary of State for the Environment and are chosen for their very wide range of relevant experience and expertise. The Commission is assisted in its works by committees of people with reputation, knowledge and experience in different spheres. Two of the most important committees relate to ancient monuments and to historic buildings respectively. These committees carry on the traditions of the Ancient Monuments Board and the Historic Buildings Council, two bodies whose work has gained them national and international reputations. Other advisory committees assist on matters such as historic gardens, education, interpretation, publication, marketing and trading and provide independent expert advice.

The Commission has a staff of over 1,000, most of whom had been serving in the Department of the Environment. They include archaeologists, architects, artists, conservators, craftsmen, draughtsmen, engineers, historians and scientists.

In short, the Commission is a body of highly skilled and dedicated people who are concerned with protecting and preserving the architectural and archaeological heritage of England, making it better known, more informative and more enjoyable to the public.

ACKNOWLEDGEMENTS

The authors gratefully acknowledge the assistance of Dr Clifford Price, Head of the Ancient Monuments Laboratory, English Heritage, in the reading of the texts.

1 AN INTRODUCTION TO METALS AND THEIR CORROSION

This chapter provides an introduction to the metals used in traditional building and sculpture in Britain. It begins with a general description of the appearance, characteristics and uses of the common metals; their properties and corrosion processes are investigated and a list of useful definitions is provided.*

1.1 METALS – DESCRIPTION AND USE

Metals fall into two categories, ferrous (containing iron) and non-ferrous (containing no appreciable proportion of iron). Non-ferrous metals are usually more corrosion-resistant than ferrous metals. Here we describe briefly two ferrous metals, iron and stainless steel, and the more common non-ferrous metals and their alloys. Many are considered in detail in later chapters.

Iron (Fe)
Iron is a dark grey metal and is the fourth most abundant element in the earth's crust. In nature it occurs in the form of iron ores which require processing to become a recognizable and usable metal. Iron is the major constituent of a range of materials including wrought iron, carburized iron (carbon steel), cast iron and steel, each of which has unique properties. (See Chapter 2 this volume, 'The repair and maintenance of cast iron and wrought iron'.)

Stainless steel
Stainless steel is steel containing both chromium (12–17%) and nickel (6–14%). Stainless steels containing about 18% chromium and from 8% to 12% nickel are

*Thanks are due to Justine Bailey of the Ancient Monuments Laboratory for her assistance in the preparation of the text.

the most widely made. They may also be called chromium-nickel steels. Stainless steels were a product of the late nineteenth-century interest in corrosion-resistant iron alloys. Their early development occurred between 1903 and 1912 simultaneously in the USA and Germany. Stainless steels can be cold worked, heat treated, cast, forged, welded, brazed and soldered. While they are not normally present in historical building fabric, their resistance to corrosion can make it appropriate to include them in a remedial works programme.

Lead (Pb)

Lead, a grey metal, is the softest, densest, and one of the most durable of the metals used in building; lead is always very pure, it is very malleable, and it has a low melting point (327°C, 621°F). The main historical use of lead in Britain has been as lead sheet. In this context it is used for roofing, guttering, pipes, flashing and solder — see Chapter 5. Lead was also used for sculpture, especially from the early eighteenth century — see Chapter 6.

Copper (Cu)

In an unoxidized form, pure copper is a salmon-red colour. It is very durable and resistant to corrosion. Pure copper is strong, ductile and malleable, which enables it to be stretched, beaten or drawn into items such as sheet and wire. It is available in three tempers: soft, half-hard and hard. In the atmosphere, over a period of 6 to 20 years copper develops a green protective patina. Run-off from copper may stain adjacent materials and it will inhibit organic growths such as lichen. See also Chapter 3, 'Traditional copper roofing'.

Copper alloys

Brass

Brass is *copper* alloyed with *zinc* (Zn). The proportion of copper can vary from 90% down to 60% or less, affecting both the properties and colour of the alloy. Other metals, such as manganese or aluminium, may sometimes be added. Brass has been used extensively for architectural castings.

See also Section 4.1 'The development of copper mining'.

Muntz metal, also known as 'yellow metal', is an alloy of 60% copper and 40% zinc. It has been used for pressed decorative roof tiles in conjunction with sheet copper on plainer surfaces.

Bronze

Bronze is copper alloyed with tin (Sn), sometimes with minor amounts of other elements too. Bronze was widely used to cast sculpture, cannon, bells, and architectural elements such as doors.

Sculpture bronze was often about 90% copper and 10% tin (see Chapter 4).

Bell metal is approximately 78% copper alloyed with 22% tin.

Phosphor bronze is an alloy of 90% copper, 8% tin with 0.2–0.5% phosphorus.

Developed in the twentieth century, this is another metal whose resistance to corrosion has led to its recent use in historic masonry remedial work.

Nickel (Ni) and nickel alloys

Nickel is a white metal which is hard and takes a high polish. It is most commonly found in buildings in the form of alloys such as Monel metal and nickel silver.

Monel metal (after 1905)

Monel metal, which is similar to platinum in colour, is an alloy of 60–70% nickel and 25–30% copper with small amounts of iron, and traces of manganese, silicon and carbon. Particularly in the USA Monel metal pioneered many of the present uses of stainless steel. Its good ductility, strength, corrosion resistance and ability to retain its properties at very high temperatures meant it was widely used there as a roofing material.

Nickel silvers

Nickel silvers are a range of alloys. Compositions are typically 63% copper, 7–30% nickel with the remainder zinc. Contrary to what the name implies they contain no silver. The colour can vary from silvery white through to pale yellow, blue, green or pink. Nickel silver was developed in China and was first imported into England in the seventeenth century. It was used to make decorative pieces such as candle holders and fire screens. By the nineteenth century nickel silver was being produced in England. From one Birmingham foundry it emerged under the name of 'Merry Metal Blanc'. Nickel silver produced in Berlin was known as 'German Silver'.

Tin (Sn)

Tin is a silvery-white metal that has been known since prehistoric times. It is soft, ductile, malleable and has a low melting point of 232°C (450°F). Because of its high resistance to corrosion its most common use now is as plating for iron and steel. Historically it was also used for items such as perforated lanterns, candle shields, wall sconces and mirror frames. Occasionally it was used for sculpture. It is a major constituent of tinmen's and plumbers' solders.

Zinc (Zn)

Zinc is a bluish-white, medium hard metal which is reasonably brittle, particularly at low temperatures. From the mid-eighteenth century it was used for roofing (see Chapter 7) and in the 1830s it began to be used for sculpture and decorative elements (see Chapter 8). Today it is widely used to provide a protective coating on steel, i.e. galvanizing, sheradizing, and zinc spraying.

Aluminium (A1)

Aluminium is a lightweight white metal with high corrosion resistance. It was rarely used in the building context until this century.

1.2.   USEFUL DEFINITIONS RELATING TO METALS

This section is based on reference 4

Metal An element which readily forms positive ions (gives up electrons). Metals are characterized by their opacity and high thermal and electrical conductivity.

Electrons Fundamental particles which are grouped around the nuclei of atoms in several possible shells, but also exist independently and produce the various electric effects observable in different materials.

Alloy A mixture of a metal and some other element(s), heated up and cooled to be a solid (the alloy).

Smelting The removal of unwanted elements in an ore to obtain a pure metal. The ore is fused with suitable fluxes to produce a melt consisting of two layers — on top a slag of the flux and gangue minerals (the portion of the ore which contains no metal), and molten impure metal below.

Flux A substance added to a solid to increase its fusability (meltability). In soldering and welding it is added to the molten metal to dissolve infusible oxide films which prevent adhesion. Alternatively it can be material added to a furnace charge to combine with constituents not wanted in the final metal produced, and which issues as a separate slag as described above.

Corrosion The gradual loss of metal solids due to chemical and/or electrochemical reactions. The process is the reversion of the metal from its unnatural refined/alloyed/smelted state to its natural state of being an ore.

Electrochemical (or electrode potential) series (Also known as the galvanic series.) The classification of elements in the order of the electrode potential which is developed when an element is immersed in a solution of molar ionic concentration.

Base metal A metal with a relatively negative electrode potential.

Noble metal Metals such as gold, silver, platinum, which have a relatively positive electrode potential, and which do not readily enter into chemical combination with non-metals.

Electrolyte A chemical, or its solution in water, which conducts current through ionization (dividing into positive ions and negative ions).

Elastic limit The limiting value of a deforming force beyond which a body does not return to its original shape or dimensions when the force is removed.

Yield point The stress at which a substantial amount of plastic (i.e. permanent) deformation takes place under constant or reduced load. In iron and annealed steels the yielding is sudden. In other metals plastic deformation begins gradually.

1.3   PROPERTIES OF METALS

The chemical and physical deterioration processes described in the following sections are controlled predominantly by the characteristics of the metal involved. The following terms are often used when describing a metal's ability to withstand corrosion processes.
(Based on reference 6, section A.)

Density is the mass per unit volume of a material and is expressed as kg/m^3. For sheet metal roofing it is usually more convenient to know the mass per unit area of the material, i.e. its weight, which is dependent on the sheet thickness.

Malleability can be considered as the ability of a metal to be formed into and over irregular shapes. Metals which are obtained in a malleable state are said to be of soft temper. Some metals become hard and brittle as they are worked and need to be annealed (see below) to restore a soft temper.

Elasticity is the ability of a metal to return to its original shape once a deforming force is removed. Every metal has an elastic limit beyond which this will not occur. Over this limit plastic deformation will occur. Slow plastic deformation is known as creep. A metal subjected to creep may eventually fracture.

Ultimate tensile strength (tenacity) is the highest load applied to a metal in the course of a tensile test, divided by the original cross-sectional area. In brittle or very tough metals it coincides with the point of fracture, but usually extension continues under a decreasing stress, after the ultimate stress has been passed.

The coefficient of linear expansion is the ratio of the expansion to the original length after a temperature rise of 1°C. Numerically the coefficiency of a metal will be very small. However, once this is applied over several degrees' temperature change to a metal sheet whose ends are restricted, the resultant lift at the centre of the sheet can be significant. Continued lifting and flattening due to changes in temperature will eventually cause the metal to crack due to fatigue.

Thermal conductivity, or the 'k' value of a metal, is a measure of the rate at which heat transmits through it. (Thermal resistance is 1/k.) Metals have high conductivity. Thermal transmittance, the 'U' value, is affected by the thickness of the sheeting and its surface characteristics. A smooth, bright surface can contribute significant resistance to the movement of heat because it acts as a reflector.

The specific heat of a material is a guide to the rate of temperature rise of a material when it is subjected to heat, and to the rate of cooling when the heat is taken away. (It is the amount of heat required to raise the temperature of a given mass of material, compared with the amount of heat required to raise the temperature of the same mass of water.)

Annealing When most metals are cold-worked they become hard because the crystal structure of the metal is deformed, elongating the grains in the direction

5

of the working and building up internal stresses. Work-hardened metals can be made soft again by annealing. The metal is re-heated to above the temperature at which recrystallization occurs. Internal stresses are released and new equi-axed grains form, resulting in softening. Grain size is time-dependent, i.e. long annealing will produce large grains. The annealing temperatures will depend on the composition of the metal or alloy, the time over which the heat is to be applied and the amount of cold-working to which the metal has been subjected.

Zinc and lead recrystallize slowly at room temperatures and do not show signs of work hardening. Zinc does, however, work best at slightly elevated temperatures.

Table 1.1 is a summary of the significant properties of the sheet metals used in historic roofing. Aluminium and stainless steel are included for comparative purposes only.

Table 1.1
Summary of the properties of metal sheet weatherings

Property	Lead	Copper		Zinc		Stainless Steel		Aluminium	
BS No	1178	2870		849		1449		1470	
Density (kg/m³)	11340	8900		7200		7900		2710	
Weight (kg/m²)	Code 3 14.17 4 20.41 5 25.40 6 28.25	0.45mm 0.60mm 0.70mm	4.00 5.40 6.30	0.6mm 0.65mm 0.70mm 0.80mm 0.90mm 1.00mm	4.32 4.68 5.04 5.76 6.48 7.20	0.38mm 0.46mm	2.96 3.58	0.70mm 0.90mm	1.897 2.439
Coefficient of linear expansion × 10 (°C −)	29.3	17.4		26.1		17.0		23	
Melting point (°C)	327°	1083°		419°		1440°		658°	
Annealing temp. (°C)		600°		100° − 150°		800°		350° − 400°	
Thermal conductivity (W/m°C)	35	384		111				226	
Specific heat (water = 1)	0.0315	0.0933		0.0935				0.2253	
Weathered colour	Grey	Green		Grey		Dull silver		Light grey	
Tensile strength (N/mm²)	12	220		280		610 − 770		80	

(Based on reference 6, page 4)

Metals corrode by reacting with their environment, attempting to return to their most stable state of a natural ore. The nature of the corrosion products of metals differ considerably. Lead forms a thin, adherent, stable patina. The patina of bronze is chemically complex. It is not usually stable when it forms in an industrial or urban environment and may even encourage further corrosion of the parent metal. Iron corrodes to form rust which is loosely adherent and many times the volume of the original iron. The process is often called 'corrosion jacking' because of the forces that can be involved. (Albert Memorial, London.)

1.4   THE DETERIORATION OF METALS

The deterioration of metals occurs as a result of physical changes, chemical changes, electrochemical changes, or a combination of these. Chemical and electrochemical changes are usually referred to as 'corrosion'.

In keeping with other materials such as masonry, once metals are placed in the atmosphere they begin to weather. For metals, the tendency is to revert to their oxide form, the form in which they occur naturally. For this and other reasons they will eventually require conservation, protection, repair or replacement in order that they and the structure they form part of can continue to function as originally intended.

The various chemical and physical processes which may be involved are defined and described in the following sections.

Corrosion processes
Corrosion is the chemical reaction between a metal and another substance. The most common reagents are oxygen (hence the name 'oxidation') and water. Corrosion is a complex process which is also affected by environmental conditions, the compositions of the metal, its position on the electrochemical series, and contact with adjacent materials, including other metals. If the layer of oxide which forms on the surface is hard, adherent and impervious, it is protective and corrosion ceases, as can happen for instance with copper and aluminium. On other metals, eg. iron, corrosion continues beneath the oxide layer, which is not adherent or impervious to moisture and air.

Chemical corrosion processes
Corrosion can attack architectural and sculptural metals in several ways. A corrosive agent may attack a metal surface evenly (uniform attack) or in selected areas (pitting). If a metal is not homogeneous throughout, certain areas of it may be attacked in preference to others (selective attack). A corrosive agent can also attack areas in a metal which were stressed during metal working (stress corrosion cracking). If a corrosion-resistant oxide layer is removed the bare metal beneath will corrode (erosion). An oxygen concentration cell is an electrolytic cell set up where oxygen is trapped between two metals or between a metal and non-metal.

Atmospheric corrosion is the most common form of corrosion experienced by metals. Moisture containing environmental gases and particles (carbon dioxide, oxygen, sulphur compounds, soot, fly ash etc.) produces chemical corrosion and/or a potential difference between two points on the metallic surface (electrochemical corrosion). Humidity level and temperature also affect the rate of corrosion. In a marine environment aerosols can deposit chloride and other salts which will accelerate the rate of atmospheric corrosion. Metals in contact with soils may pick up other corrosive agents.

Electrochemical (galvanic) corrosion
Electrochemical corrosion and galvanic corrosion are the terms which relate to

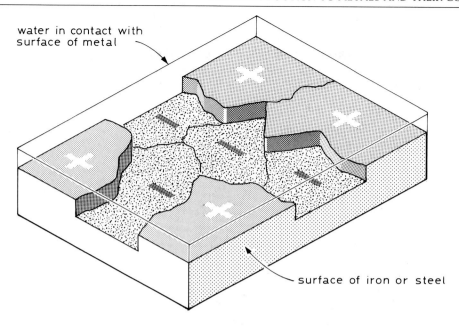

water in contact with surface of metal

surface of iron or steel

at anode	$Fe \rightarrow Fe^{++} + 2e$
at cathode	$O_2 + 2H_2O + 4e \rightarrow 4OH^-$
combined reaction	$4Fe + 3O_3 + 2H_2O = 2Fe_2O_3H_2O$
	iron oxygen water rust

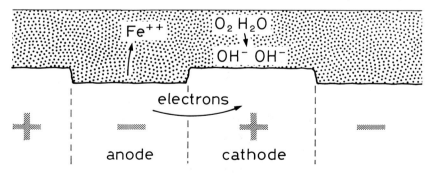

Corrosion of iron and steel occurs in the simultaneous presence of water and oxygen

after British Steel Corporation / I.McC.

Figure 1.1 **Electrochemical corrosion of iron**
The formation of rust on iron is the result of two electrochemical reactions which require the presence of both oxygen and water. Surface treatments on iron, such as oil, wax and paint, have been applied to isolate the iron and prevent this reaction.

Table 1.2
The electrochemical series

Metal	Chemical symbol	Electrode potential (volts)	
Silver	Ag	+ 0.80	More noble*. Less reactive.
Copper	Cu	+ 0.35	Cathodic, protected end.
Hydrogen	H	0.00	
Lead	Pb	− 0.12	
Tin	Sn	− 0.14	
Nickel	Ni	− 0.23	
Iron	Fe	− 0.44	
Chromium	Cr	− 0.56	
Zinc	Zn	− 0.76	
Aluminium	Al	− 1.00	Less noble*. More reactive.
Magnesium	Mg	− 2.00	Anodic, corroded end.

*There are two different sign conventions relating to anodes and cathodes. The one selected here denotes cathodes as positive and anodes as negative, as this is the convention most commonly used in references relating to the nature of building materials.

the increased corrosion of one metal due to its contact with another or, in certain instances, the same metal.

The corrosion resistance of a metal is determined by its position in the electro-chemical series in which metals are arranged according to their electrical potential (see Table 1.2). The tendency of a metal to corrode is greatly increased when it is in contact with another metal lower down the scale, by means of a conducting liquid known as an electrolyte. The electrolyte enables a flow of electrons from the less noble metal to the more noble. The less noble metal behaves as the anode and suffers increased corrosion, while the more noble metal will behave as a cathode and be corroded less than when uncoupled.

For electrochemical corrosion to take place there must be an anode (negatively charged area), a cathode (positively charged area) and an electrolyte (conducting medium) (see Figure 1.1). The electrolyte may be rainwater or condensation. It may be an acid, an alkali or a salt. The formation of an anode and a cathode may even occur within one metal due to the presence of impurities, differences in work hardening or local differences of oxygen concentration on the surface. The latter may occur where two pieces of the same metal are in close proximity so as to trap a film of water between them ('crevice corrosion').

The rate of additional corrosion depends on a number of factors such as the size of potential difference between the metals, temperatures and composition of the electrolyte. The relative size of each metal is also significant because the severity of galvanic corrosion depends on the rate of flow of electrons between the two metals. If the more noble (protected) metal is much larger than the less noble (corroded) metal, the deterioration of the less noble metal will be more rapid and severe. On the other hand, if the more noble (protected) metal is much smaller than the less noble (corroded) metal, the deterioration of the less noble metal will be considerably less significant. For example, iron rivets fixed in a large copper sheet will deteriorate rapidly, while if a large iron sheet is fixed with copper rivets the iron plates will deteriorate slowly, all other environmental factors being equal.

When two dissimilar metals are embedded in good electrical contact in wood the less noble will corrode and the more noble will be protected, e.g. an iron fixing would corrode sacrificially in relation to a copper one. A fastener of one metal may corrode electrochemically if its exposed head receives more oxygen than its embedded tail or if two timbers with different acid or salt contents are being joined. (See BRE Digest 301, *Corrosion of Metals by Wood*, September, 1985.)

Mechanical deterioration processes

There are several purely physical processes which will cause the deterioration of metal.

Abrasion will cause removal of the metal surface which may be protective. Lead flashing to valleys on slate roofs can be eroded by loose pieces of slate which score the lead surface.

Fatigue is the failure of metal which has been repeatedly stressed beyond its elastic limit. It occurs in copper, lead and stainless steel roofing where the necessary allowances for thermal expansion and contraction have not been made.

Creep is the permanent distortion of a soft metal which has stretched due to its self-weight. Thin areas of the metal will be among the first to fail. Creep can be found in lead sculpture which has an inadequate or corroded internal armature.

Heat, usually in the form of fire, will cause many metals to become plastic, distort and fail. Others, such as cast iron, have higher melting points and hence good fire resistance.

Distortion, permanent deformation or failure may occur when a metal is *overloaded* beyond its yield point. This can occur for a number of reasons, eg. increased live loads or dead loads, thermal stresses, or structural modifications altering a stress regime.

Connection failure

Both chemical and mechanical processes can cause the breakdown or reduced effectiveness of structural metal fixings such as bolts, rivets and pins. Stress failure is often a contributor in a breakdown situation. Iron connections which are water traps are particularly susceptible.

REFERENCES

1 Everett, Alan, *Mitchell's Building Construction – Materials*, B. T. Batsford Ltd, London, 1970.
2 Brewer, C W, 'Ancient Methods of Metal Fabrication', *The Conservation and Restoration of Metals*, Scottish Society for Conservation and Restoration, proceedings of the symposium, Edinburgh, 1979, pp 1–9.

3   British Steel Corporation, *The Prevention of Corrosion on Structural Steels*, Technical pamphlet, BSC Sections, Cleveland, UK.

4   Collocott, T C and Dobson, A B (eds), *Chambers Dictionary of Science and Construction*, W & R Chambers, revised edition, 1976.

5   Gayle, Margot, Look, David W and Waite, John G, *Metals in America's Historic Buildings — Uses and Preservation Treatments*, US Department of the Interior, US Government Printing Office, 1980.

6   Institute of Plumbing, *Sheet Roofing Data Book and Design Guide*, Technical Committee; Oxford District Council and the Institute of Plumbing, Oxford, 1978.

See also the Technical Bibliography, Volume 5.

2 THE REPAIR AND MAINTENANCE OF CAST IRON AND WROUGHT IRON

This chapter was written in collaboration with Mr Geoffrey Wallis whose assistance is gratefully acknowledged.

2.1 THE DEVELOPMENT AND PRODUCTION OF IRON

The forms of iron

Iron is the fourth most abundant element in the earth's crust. Iron does not occur in nature as a metal, but as ores, which require processing before they become usable metal. This chapter deals with the developments in processing and working the metal in Britain, and the various forms of iron which developed. In order of appearance these were wrought iron, carbon steel (carburized iron), cast iron and mild (bulk) steel. The forms of iron have distinctive characteristics originating from their method of manufacture and carbon content. Wrought iron is the purest form of iron followed by the steels, cast iron having the highest carbon content. Carbon steel was made by 'adding' carbon to wrought iron and is the traditional form. Mild steel is made by 'removing' carbon from cast iron and is a modern form of iron.

Wrought iron

The exact date of the commencement of iron production is not known, but over 4000 years ago wrought iron was being made in the Middle East. Ironwork techniques spread across Europe and, by about 450 BC, reached Britain, where the primitive beginnings of iron making had begun some 300 years earlier.

The first simple process used for making iron at the time was called *direct*

Wrought iron was used prolifically in gates and railings, especially during the seventeenth and eighteenth centuries. Its malleability enabled it to be worked, by a wide range of techniques, into many different forms. Square or rectangular bar was connected by a series of carpentry-style joints such as mortice and tenon, halving and pinning, to provide the main structure of a gate or railing. Decorative elements such as leaves and scrolls were applied to this, often attached by the process known as heat or forge welding. Paint and gilding were common decorative finishes. It should not be assumed that the ubiquitous black paint now seen on many wrought iron structures of this period is the original colour. (Tijou Screen, Hampton Court Palace)

14

reduction. Iron oxide was heated in contact with carbon (charcoal or coke), so the carbon and the oxygen of the ore combined, leaving iron behind.

$$Fe_2O_3 \quad + 3C \quad \rightarrow 2Fe \quad + 3CO$$

Iron ore   Carbon   Iron   Carbon monoxide gas driven off.
metal

The process was carried out in very small furnaces called bloomeries, a 'bloom' being the small lump of iron produced. The earliest form was the low bloomery, a bowl-shaped hole in the ground lined with clay. The later high bloomery was a clay shaft above ground level. After smelting, the bloom was removed and forged to further reduce carbon and other impurities.

Ironmaking remained a village craft rather than an industry until a new process was introduced. This was made possible by the blast furnace developed in Belgium circa 1400. In the *indirect reduction* process, the manufacture of wrought iron occurred in two stages. Iron ore was first smelted in blast furnaces to produce 'pig iron', which was in effect cast iron. The cast iron 'pigs' were then converted in the finery to wrought iron by the removal of carbon (decarburization). Details of this process are described in the following section.

Blast furnaces developed and increased in capacity. They needed to be established near a source of water power, iron ore and charcoal. The Sussex Weald had all of these and by 1700 was Britain's biggest iron manufactury.

A severe shortage of charcoal for the first stage of the indirect process led Abraham Darby in 1709 to use coke, a modified form of coal. Darby, one of the first specialist ironfounders, established the Coalbrookdale Company at Coalbrookdale, Shropshire in that year. The continuing charcoal shortage for the conversion of pig iron to wrought iron was resolved in 1784 by Henry Cort (at Fontley, Hants), who kept the iron and coal separate and developed the process known as 'puddling'. In this, 5-cwt batches of iron were reduced in an indirect or reverberatory furnace.

Cast iron

Although cast iron has been made for some time as part of the indirect wrought iron process its widespread use did not develop until the sixteenth century. The breakthrough occurred in 1794 with Wilkinson's invention of the cupola, a small blast furnace used for remelting pig iron rather than smelting ore. It provided cast iron by a quick, relatively economical, simple process and was therefore the impetus for the establishment of a multitude of small foundries. By 1800 the Industrial Revolution was well under way and the production of castings increased for the rapidly developing engineering industry.

Steel/carbon steel/carburized iron

Traditional steels have a combined carbon content of about 0.5% to 1.5%. This gave traditional steel the important ability to be hardened by quenching in water after being heated to a bright red heat, an ability which the purer (wrought) iron

15

did not have (reference 2, pp 1—3). The different characteristics of pure iron and steel were long understood, but the ferrous metallurgy of this difference was not fully exposed until well into this century. Because of its high carbon, traditional steel is also referred to as 'carbon steel' and 'carburized iron'.

The production of traditional steel was achieved in several ways. Prior to the development of the blast furnace, conditions in the bloomery furnace were modified so the wrought iron bloom retained some carbon. Alternatively, bloomery iron was heated on a bed of charcoal which added carbon to its outer surface and thus hardened it. Following the development of the blast furnace and the production of cast iron as the first stage, several more production routes for carbon steel were opened up which essentially involved removal of a portion of the carbon in blast furnace iron. The first of these methods produced carbon steel in a bloom, i.e. the Cementation Process, 1772. Later methods produced liquid steel for casting, e.g. Huntsman Steel, Crucible Steel and eventually Bessemer Steel.

Details of these processes and the history of steel making are covered most thoroughly by K. C. Barraclough, references 2 and 3.

The situation in 1850

In 1850 wrought iron production was at its peak but the material was beginning to be considered expensive. Cast iron was finding increasing use as engineering continued to expand. Carbon steel by the cementation/crucible route was slow and expensive to produce, but it was the only form suitable for hardening and tempering and so was in demand for tools, where it was selectively used for the cutting edge alone.

Mild steel/bulk steel

In 1856 Henry Bessemer invented the Bessemer Converter in an attempt to make wrought iron, or something like it, on a large and economic scale. Wrought iron needed a very labour- and fuel-intensive process which resisted all attempts to mechanize it. Bessemer succeeded in producing a metal which was similar but physically different to wrought iron, what is now called mild steel.

The production of steel escalated after 1880 and wrought iron soon lost its pre-eminence. The Atlas Forge of Thomas Walmsley, established at Bolton, Lancashire, was the last mill in Britain to make wrought iron. Operation ceased in 1973 and much of the equipment has been re-erected at Blists Hill Open Air Museum, Ironbridge, where it is hoped to recommence wrought iron production on a limited scale.

2.2 METHODS OF PRODUCTION

Wrought iron production by the indirect method

The first stage in the production of wrought iron was the preparation of puddled iron. Pig iron was smelted in the furnace, subjected to a reducing condition when it apparently 'boiled'. During boiling the iron was worked continuously by the puddler and as more impurities were removed the iron became stiffer. The boiling stage ended when carbon monoxide no longer bubbled through the iron, virtually no carbon being left. The pasty mass of iron was then formed into balls. Although some quite complex chemical reactions took place during the making of puddled iron, the process, unlike steel manufacture, was not scientifically controlled, but depended very largely on the skill and experience of the man in charge.

The second stage of wrought iron production began at the shingling hammer, where the iron balls were hammered to expel surplus slag or cinder (shingled). Shingling was completed in minutes and the finished product was a bloom approximately 5 inches × 5 inches × 3 feet. The bloom, still at bright red heat, was then passed through rolling mills, becoming more enlongated and thinner in section after each pass, and finished as puddled iron bar.

The mechanical strength of puddled iron was low. Improvements in both ductility and tensile strength were made by further re-heating and re-working, hence the term 'wrought'. The first rolling produced 'Crown iron', the lowest but most commonly used grade. Additional heating and rolling produced the better grades Crown Best, Best Best, Best Best Best and Treble Best. The bar was also rolled into sections such as rounds, flats, squares, half rounds, angles, tees, bridge rails, pit rails and many more.

The standard sections of wrought iron at the end of the eighteenth century were bar iron, angle and T irons, channel iron (half H iron), rolled girder iron (rolled joist iron, beam iron, I iron or H iron), various special sections (sash bar, beading iron, cross iron, quadrant iron), iron bars, rivet iron, chain iron, horseshoe iron, nail iron, plate iron, coated iron (e.g. with tin or lead), corrugated sheet iron (generally galvanized) (reference 20).

The quality of wrought iron can be variable. There can also be considerable variation in section in early iron which was wrought by hand before the introduction of rollers. Heat welds within a section may not be properly formed and may start to pull apart.

Cast iron production
There are four stages to the production of new castings.

Pattern making
This is the production of an original from which one or more castings are made. Patterns are nowadays often constructed of timber, g.r.p. (glass reinforced plastic), wax, plaster or another metal, and may be complex assemblies with several parts. They are frequently produced outside the foundry by specialist pattern-makers, but for simple components an original casting may serve as a pattern. The pattern should be slightly over-size to allow for contraction, and constructed to allow it to be drawn from its mould.

Moulding
The traditional material of which moulds were made was a damp sand clay mixture called 'green sand', which is still commonly used today.

Molten metal was poured into a void formed by the pattern as illustrated in

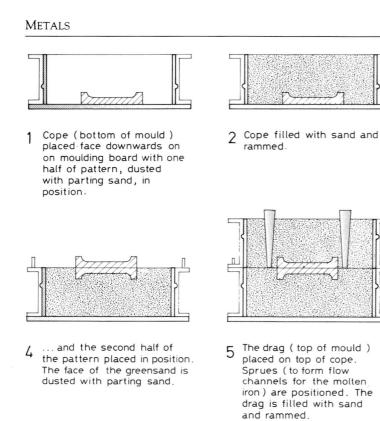

1 Cope (bottom of mould) placed face downwards on on moulding board with one half of pattern, dusted with parting sand, in position.

2 Cope filled with sand and rammed.

3 Mould board placed on top and the cope inverted ...

4 ... and the second half of the pattern placed in position. The face of the greensand is dusted with parting sand.

5 The drag (top of mould) placed on top of cope. Sprues (to form flow channels for the molten iron) are positioned. The drag is filled with sand and rammed.

6 The sprues are removed and a mould board is placed on top of the drag. The cope and drag are separated and the drag inverted.

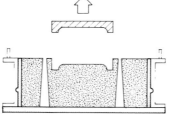

flow gates

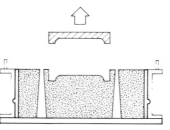

7 The pattern is removed from the cope and flow gates cut in.

8 The half-pattern is removed from the drag.

9 The cope and drag are reassembled; pouring of molten iron may now commence.

Figure 2.1 General procedure for mould-making in green sand
This traditional technique of making cast iron is still readily available. The replacement of cast iron with cast aluminium should be strenuously avoided.

Figure 2.1, which shows the technique for moulding a simple, small casting without cores. For large castings the green sand bed of the foundry floor was often used, and for flat items, such as floor plates and firebacks open moulds

The capital to this gate pier is a jigsaw of cast iron sections bolted and fitted together. Its design reflects the heaviness and repetitive nature that is often seen in cast iron items. As cast iron is crystalline and brittle, excessive corrosion, impact or heat stress can cause it to fracture. Cast iron shrinks about 1% on cooling from its molten state, which may be of no consequence on smaller, decorative pieces but will certainly matter on items such as the 9 feet high structural pier column supporting this capital. (Coalbrookdale Gate, Kensington Gardens)

were used. In these the upper surface of the metal was open to the air, and took on an undulating surface.

Cores were traditionally of baked sand. Nowadays sand bound with sodium silicate and cured with CO_2 gas is commonly used.

Casting

Traditionally pig iron, cast iron scrap and limestone flux were melted in a coke-fired cupola. Today gas, oil or electrically fired crucible furnaces are commonly used. Sufficient charge is melted and withdrawn from the furnace. Slag is removed, and the waiting green sand moulds are filled in one pouring. Once cooled, castings are broken out and passed to the fettling shop.

Fettling

This is the final stage, in which the casting's pouring gates and risers are removed, the mould line or 'flash' is dressed off and the raw casting is brushed clean.

2.3   PROPERTIES AND USES OF CAST IRON, WROUGHT IRON AND STEEL

Wrought iron − properties and development

- Wrought iron is the purest form of iron, with less than 1% carbon.
- It has a high slag content of iron silicate of between 1% and 4%. This exists in a purely physical association, i.e. it is not alloyed as it is in steel.
- The structure of wrought iron is fibrous and it is able to withstand tension well.
- It is relatively good at resisting corrosion.
- It is relatively soft and malleable but can also be tough and fatigue-resistant.
- It can easily be worked by forging, rolling and bending. It can be worked hot or cold.
- Wrought iron can be heat welded, i.e. if two pieces are brought to a white heat and forced together they will fuse.

Up to the fourteenth century wrought iron was mainly used for spears, arrow points, swords and knives. From the fourteenth to the eighteenth centuries significant uses for wrought iron were agricultural implements and on church doors as decorative and semistructural strapwork, hinges and nails. Structurally its use was restricted to tie bars in medieval churches and cathedrals. Its decorative use also became important for church screens, gates and railings.

From the eighteenth century on, wrought iron was used structurally for beams and girders which utilized its good strength especially in tension, often built up from plate and angles riveted together. Following technological advances in rolling machinery, in about 1840 rolled beams became available. Often these were strengthened by flange plates and web stiffeners held in place by rivets. During this period wrought iron was frequently used in conjunction with timber and masonry construction. The tensile strength of wrought iron led to its use as cables and links for suspension bridges, as tie bars and chains.

Cast iron — properties and development

- Cast iron is an alloy of iron and carbon. It has a relatively high carbon content of up to 5%. The most common traditional form is grey cast iron.
- Common or grey cast iron is easily cast but it cannot be forged or worked mechanically either hot or cold.
- Its structure is crystalline and relatively brittle and weak in tension. Cast iron members fracture under excessive tensile loading with little prior distortion. Cast iron is, however, very good in compression.
- Cast iron generally has good fire resistance and good corrosion resistance.
- In grey cast iron the carbon content is in the form of flakes distributed throughout the metal. In white cast iron the carbon content is combined chemically as carbide of iron (Fe_3C). White cast iron has superior tensile strength and malleability. It is also known as 'malleable' or 'spheroidal graphite' iron.

Cast iron began to be used structurally as columns and beams at the end of the eighteenth century. Columns were generally hollow and commonly circular, hexagonal or octagonal in shape, and often plain fluted, or beaded. Beams were shaped and proportioned in recognition of cast iron's weakness in tension, with greater thickness in the bottom flange than in the top. Beams were often I or inverted T section, sometimes adopting the structurally efficient 'fish belly' profile on elevation.

Inside radii were usually generous, partly to improve molten metal flow and partly to reduce stress concentration. This also helped accommodate differential cooling stresses. In contrast, the external corners of cast iron sections are frequently sharper than those of wrought iron or steel.

Connections between cast iron sections were usually simple sockets, spigots and/or wrought iron bolts.

Steel — properties

- Steel is another alloy of iron and carbon. Its composition can range from mild steel, which has a carbon content of 0.2%, to the complex modern alloys which contain considerable amounts of other elements.
- In mild steel carbon is alloyed to the iron. Where the carbon content is high enough, steel can be heat treated, whereas the low combined carbon content of wrought iron does not allow this.
- Although there is a lack of conclusive scientific data, the corrosion resistance of steel is generally thought to be inferior to that of wrought iron.
- Steel can be fusion welded (as can wrought iron and cast iron), but it is relatively difficult to join two pieces of steel reliably by forge welding.

Cast iron is particularly suited for use in elements subjected to compression loads, but, unlike wrought iron, it has little ability to withstand tension forces. In structures such as the Palm House at Kew Gardens (c.1814) cast iron and wrought iron were intermixed according to their structural abilities. Compression and decorative members such as columns, brackets, stair treads and 'balusters' were of cast iron. The glazing mullions, arches and tension rods were made of wrought iron.

2.4   DISTINGUISHING THE FORMS OF IRON

Distinguishing wrought iron from cast iron

- The type, design and structural function of members is often the best method of identification. Wrought iron was generally used in tension while cast iron members were primarily used in compression. Composite structures of both irons usually reflect this but further confirmation should always be sought, as this was not always the case.
- Cast iron elements are usually more massive in appearance and were often repetitive.
- The mould line or 'flashing' may still be visible on cast iron pieces.
- Wrought iron members were often built up from riveted sections. Fire/forge welding is a joining method common on wrought iron but never used on cast iron.
- Wrought iron may have a rolled or hand beaten surface. The surface of a

22

cast iron member will often reveal blow holes, casting flaws or inclusions.
- It is relatively easy to pare away a sliver of wrought iron with a sharp cold chisel. Cast iron, by contrast, chips away.
- The crystalline structure of cast iron and the fibrous structure of wrought iron can be seen at fractures.

Distinguishing wrought iron from steel

- The date of construction is often a clear indication. Steel was not in common usage until the 1880s.
- Construction and fixing techniques can be indicative. Modern friction grip bolts and original fusion welding are exclusive to steel. Forge welding is a usual jointing method for wrought iron. Wrought iron jointing practice is often similar to that of timber construction.
- A metallurgical check will show whether the carbon is in physical association (wrought iron) or alloyed (steel), and its percentage content.
- Metal workers can differentiate between wrought iron and steel by the colour and shape of sparks when grinding or by inspecting bend fractures. The fibrous nature of wrought iron and the crystalline nature of steel may be revealed either by acid etching or by testing ductility, i.e. percentage elongation at fracture.

Testing iron

On the rare occasion when *in situ* analysis and other research do not provide a satisfactory differentiation, samples can be sent to laboratories such as The Welding Institute and the British Cast Iron Research Association. When testing is being contemplated the testing organization should be contacted initially to determine the type of results which will be provided and the size and number of sample/s required. Samples of iron should be taken from lightly loaded areas and in minimal quantities.

2.5   DETERIORATION OF IRON

Essential maintenance

In the presence of water and oxygen, iron will corrode. Essential maintenance therefore centres around keeping these two away from iron surfaces. Iron's tendency to corrode makes it essential that maintenance is carried out frequently, regularly and competently.

All rainwater management arrangements should be kept functioning properly. A build-up of water in a cast iron downpipe can freeze and cause a serious fracture. All joints should be inspected periodically. Minor gaps can be filled temporarily to keep water out, although this should never be considered anything more than a short-term remedy. Linseed oil putty overdries and loses its adhesion. It should be inspected annually and renewed every few years. Small defects in paintwork should be repaired as soon as they are noticed so as to curtail development of corrosion under the paint surface.

Voids in cast iron elements exposed to rainwater should not be filled with concrete. The shrinkage of concrete during curing can leave a crevice which can retain water, offering prime conditions for rusting and giving the swelling rust a surface to push against.

Structural failure

Cast iron members which are repeatedly stressed beyond their (low) elastic limit will fracture. Wrought iron is a more forgiving material and will deform, usually quite considerably, when stressed within its elastic limit. Wrought iron fixings may continue to function effectively even if deformed by overloading in the past. However, brittle fracture of wrought iron can develop from either a defect in the wrought iron or a point of stress concentration.

Defects integral to the material

Mechanical defects in the manufacture of cast iron can reduce the mechanical strength of a member but are rarely big enough to do so significantly. Casting flaws were common during the nineteenth century when founding quality was poor. Foundries often attempted to conceal these with materials such as Beaumont's Egg (a mixture of wax and iron filings).

A wrought iron member may delaminate partly or completely at the point where pieces have been forge welded together incompletely or between webs and flanges of beams.

Indiscriminate alterations

Repairs of the past which were made without an understanding of the cause of failure often give rise to failures elsewhere by transferring stresses. Crude repairs can be obtrusive and should be remade.

The piercing of iron structural members for new services should be avoided.

Graphitization of cast iron

This is a form of corrosion which affects grey cast iron by removing the ferrous matrix. It is a form of anaerobic corrosion (occurring without oxygen) in acidic soils or water, especially seawater. The different phases of cast iron (graphite, iron, silicon etc.) act as electrodes in a corrosion cell and the matrix is substantially dissolved. The change in dimension and surface texture accompanying graphitization is virtually nil, but the mechanical strength of the layer is negligible and it can therefore be a hidden danger. A graphitized area can show up as a black surface which may be slightly blistered. The surface can often be broken up easily with a knife blade and the underlying material will crumble. Graphitization of cast iron is not common but should be sought where conditions make it a possibility.

Embrittlement of cast iron

Cast iron elements may embrittle with age. It is thought that this might relate to excess phosphorus in the casting metal. Brittleness with hardness may also be caused by chilling during casting. In handling cast iron, care should be exercised at all times.

Corrosion

The corrosion of iron is the formation of iron oxide (rust) by the reaction of iron with oxygen and water. Chemically, rust is very similar to the minerals from which the iron was originally extracted.

For iron to corrode it must have the simultaneous presence of oxygen and water. In the absence of either corrosion does not occur (except for graphitization). Corrosion prevention usually involves the application of a coating to separate the iron from the water and/or oxygen in its environment. Prevention of water penetration and retention is, therefore, a vital aspect of corrosion prevention.

An increased amount of corrosion will occur when a difference of electrical potential arises between two points on a wet iron surface. This can be due to contact with a different metal, between different crystals in the iron (e.g. cementite and ferrite), between different qualities or conditions (annealed and work hardened), and between areas of different aeration in the moisture layer of film.

Further details of the corrosion processes of metals may be found in Chapter 1, this volume.

Corrosion also does not occur if the basic electrochemical reactions are suppressed. This can be achieved by the use of corrosion inhibitors and the application of cathodic protection (zinc coating, galvanizing, aluminium coating, anodizing, cadmium plating etc.).

The application of paint has long been recognized as the most practical method of protecting iron structures. It is important to maintain a continuous paint layer to ensure proper metal protection. When corrosion starts attacking the coated metal at a defective point in the coating, it tends to propagate under the coating itself because the areas which have less access to oxygen are differently charged from the exposed ones. The corrosion will therefore spread under the coating.

Coatings are applied in several layers to reduce the chance that pin holes and thin areas in these will coincide in all layers. The aim is to make the isolation of the iron from its environment as complete as possible.

2.6   REPAIRING CAST IRON AND WROUGHT IRON

As with all other historic fabric, ironwork will have a historic importance which must be identified at an early stage. The item's merit, in terms of age, uniqueness of design, materials, size, technological development, association with persons or events, exceptional workmanship or design qualities, must be understood before decisions regarding repair, maintenance and preservation can be made.

The principal aim of any work must be to halt the processes of deterioration and stabilize the item's condition. Repair is a second option which becomes necessary only where preservation is not sufficient to ensure mid- to long-term survival. Repair should always be based on the fundamental principle of *minimal disturbance*.

The following are good practices which arise from this principle:

- Retention of as much existing material as possible; repairing and consolidating rather than renewing.
- The use of reversible processes wherever possible.
- The use of additional material or structure to reinforce, strengthen, prop, tie and support.
- The use of traditional materials and techniques. New work should be distinguishable to the trained eye, on close inspection, from the old.
- The item should be recorded before, during and after the work.

Assessing the problem

When assessing a cast iron or wrought iron structure it is important to consider the following:

- How the structure was put together and how it was intended to work structurally when designed.
- Which members are primary, i.e. critical in load bearing, and which are secondary or decorative.
- How the structure is working now. This is not necessarily how it was working when built, so the tensile, compressive and compound stresses within the members shoud be determined.
- The condition of the structure as a whole and the condition of the individual parts.
- Vulnerable parts such as footing, fastenings, interlocking parts and water traps. These should be inspected most carefully. Paint may need to be removed for a proper inspection.
- Bolts and rivets. These are particularly vulnerable and should be investigated thoroughly. While the head may be in good condition the shank may be virtually corroded away. Crevice corrosion may have occurred between bolted or riveted faces.
- The taking of levels to check for movement can be very valuable in showing up areas of a structure which are currently most stressed. If movement has occurred it is essential to determine how the structure has accommodated this. Castings may have broken, fastenings failed or loads on adjacent components increased.

The solutions for cast iron and/or wrought iron structures should only be planned when a structure is understood thoroughly. This may require engaging a structural engineer with experience in these materials and structures.

Structural repair options

Leave alone

If a structure has reached a new equilibrium and its members are not stressed beyond their capacity it is advisable not to interfere with the structure, provided its metalwork surfaces can be stabilized *in situ*.

Reduce loads

This includes reducing spans, doubling beams, introducing intermediate columns, lightening dead weights and live or applied loads. Several of these options will have visual implications which must be carefully considered in the conservation context.

Change stress regime

This may include introducing tension rods to reduce tension stresses in cast iron, or removing structure added to the original which has complicated the distribution of loads.

Strengthen/reinforce

Members may need to be strengthened by the addition of plates, doublers or strengtheners. Interior columns may be strengthened by filling with a material of good adhesion (e.g. resin) if never exposed to moisture.

Dismantle and re-erect

Cast iron and wrought iron structures are usually not difficult to dismantle and re-erect. The work should be undertaken with care and not by inexperienced persons. The procedure has several advantages. A thorough understanding of the behaviour and condition of the structure can be gained, original components can be used for re-casting, rusted joints can be cleaned properly and new fixings inserted. Cleaning and painting can be carried out more effectively, and defective parts are more likely to be revealed. Every attempt should be made to dismantle in reverse order to construction. Every piece should be marked in at least two places. Dismantling should be carried out only as far as necessary, and care taken to ensure that the remaining structure remains stable.

In situ repair techniques

Cast iron and wrought iron fabric is of intrinsic value and it is preferable to repair rather than renew. Several methods of *in situ* repair are available.

Welding cast iron

While welding of cast iron is possible with great expertise and careful supervision, it is not always possible to be sure of the integrity of the repair. During welding the metal becomes very hot and can undergo tremendous thermal shock; it may recrystallize if excessive heat is applied. A good quality weld in cast iron usually requires removal of the section to a workshop where it can be preheated before welding and postheated after welding to ensure a gradual temperaure change within the metal. It is inadvisable to attempt to weld large sections of cast iron on site.

The following welding processes are commonly used for cast iron:

Type	Heat source	Shield (from O_2)	Process
Fusion	Electric arc 3,000°C	Flux Inert gas	'Stick' weld MIG, TIG
Metallic bond	Fuel gas + oxygen 900°C	Flux	Brazing

(G. Wallis)

In a large section of cast iron a fusion (arc) weld of high integrity is especially difficult to achieve, so caution is advised. Metallic bond (gas) welding is a more reliable technique. A far lower temperature is used and heat is applied and removed at a slower rate.

Successful welding of cast iron can be a relatively expensive operation and so whenever possible a cold repair is recommended.

Welding wrought iron

Most grades of wrought iron can be welded satisfactorily. The Welding Institute can do tests on iron samples to determine the best welding parameters and filler rod.

Butt welds are usually recommended for structural joints as this ensures that all sections of any laminations which may be present will be attached. Fillet welds can be used provided they incorporate a sound edge of wrought section.

Rivets should not be replaced by welded joints. It is best not to weld near rivets as welding distortion may stress these and form a gap between the joined materials. This destroys the face friction in the joint and allows water to enter by capillary action.

Wrought iron can be successfully welded to steel and stainless steel. Bimetallic corrosion is possible in this situation, although there is a dearth of practical evidence for this.

Cold repair techniques — castings

Fractures in cast iron can be repaired or stabilized by several 'cold' methods. Stainless steel or non-ferrous metals should be used whenever possible. Cold repair techniques include:

- Straps. These can often be hidden. The plate should be bedded on a suitable medium to prevent a water trap.
- Threaded studs, screwed into both sides of a fracture.
- Dowels or plain pins, with one or both ends threaded and/or glued into prepared recesses.

The traditonal cold repair method was to insert a 'dumb bell' shaped piece of wrought iron across a fracture. This method has developed into several contemporary patent cold stitching systems. They produce a sound repair to fractures, are strong and are easy to use on site.

Once the fracture is realigned, groups of holes are drilled across it and linked by pneumatic chisel to form a series of slots. Locks of work-hardening nickel alloy are then driven in. Holes are drilled along the line of the fracture between these, then tapped and filled with studs, each stud interlocking with its neighbour. All excess metal is then sheared off and the surface is ground and painted.

Metal stitching can be used on all cast iron which is over ¼ inch (6 mm) thick. A notable benefit of the system is that virtually no heat is introduced into the metal. Metal stitching is not applicable to wrought iron, although it is suitable for other cast metals such as bronze. It can be used between cast iron and metal of similar hardness, such as steel.

The use of fillers

It is usually preferable to fill a wasted but otherwise serviceable component than to renew it. There are several good quality metal replacement materials on the market based on steel particles with an epoxy resin binder. Epoxy resin alone has a high thermal expansion which is not compatible with that of iron. If used without a high proportion of metal filler such as in a thick layer of adhesive, problems with differential expansion may occur. Metal fillers may be used on cast iron and wrought iron to make good an area of corrosion or defects such as casting flaws, and water traps, and where the member is still structurally sufficient. The area must be prepared properly to receive the material.

New castings

Seriously corroded, broken or missing castings may need to be recast. Grey cast iron should be replaced in the same material. Replacements in cast aluminium are not historically accurate, may corrode sacrificially to cast iron and are weaker. The processes involved in production of a new casting have been described in Section 2.2.

Cast iron shrinks about 1% on cooling from melting point. Where this is acceptable, existing pieces may be used as patterns. Where shrinkage cannot be tolerated or where the shape of the item does not permit direct moulding, a new pattern will need to be made. For an original item to be used as a pattern, it must be able to be drawn out of each side of the mould. Pieces with re-entrant angles, therefore, cannot be used without modification.

It is always advisable to speak to a traditional iron foundry early on in a job to determine the work and cost involved in a new casting.

Replacement wrought iron

For many years the only source of wrought iron has been recycled material. While it may cost more to re-roll old wrought iron for a replacement unit than to use steel as a substitute, historical authenticity makes this worth while. The Thomas Walmsley forge from Bolton, Lancashire, relocated at Blists Hill, Ironbridge, is

The paint on ironwork must be continuous to be effective. Corrosion begins at breaks in the surface and then spreads beneath it. Paintwork on iron should be maintained annually. Defects should be made good by completely removing all rust and priming and painting before any more forms. Painting over rust is a waste of time and money. Ironwork which is coated with many layers of paint will lose much of its crispness and detail. If such a coating is defective, it is sensible to remove and replace it. On structures of historical importance this should not be done until layer samples have been taken and analysed microscopically to determine the original colour and the disposition of gilding. The cleaning of ironwork should not damage or rework its surfaces. Cast iron and wrought iron require different methods of cleaning. A test area which includes the various sections and forms of iron should be selected and the cleaning method/s demonstrated on this so that the correct variables for the job can be determined. The use of experienced operatives is essential. Surface preparation of the iron must be thorough for the selected paint system to be able to perform properly.

to begin producing wrought iron in the near future. It is hoped this will provide a second source in the UK additional to the recycled material.

Rehousing ironwork into masonry

The ends of all existing ironwork which is to be reset into masonry must be either cleaned and treated thoroughly or replaced. If they are to be reclaimed abrasive cleaning will probably be necessary to remove completely all corrosion prior to painting with epoxy paints and fixing with lead or lead wool packing. Severely corroded ends can be tipped with stainless steel, Delta bronze or new wrought iron for the distance of the housing plus at least 12 mm from the masonry face. For bronze a lapped joint is formed and bolted with a bed of sealant between the bronze and iron to prevent water penetration. In the case of stainless steel a welded butt joint is formed. Bimetallic corrosion is possible with all the non-ferrous materials, but practice has yet to confirm this.

2.7 CLEANING AND SURFACE PREPARATION

The function of traditional paint on iron

Cast iron and wrought iron were painted primarily for protection against corrosion and secondarily for decorative effect. Each coat of paint had specific and different purposes.

The *primary coat* could include any of the following:

Red lead, a bright orange oxide of lead, was also known as minium or orange mineral. Red lead has been in use as a protective pigment for over 2000 years and has the effect of inhibiting the rust-forming reaction. Removal or disturbance of this coating has several health and environmental implications.

Iron oxide, the natural and most stable form of iron, is a reddish or brownish pigment. Although it was ubiquitous in nineteenth-century primers, iron oxide had little rust-inhibiting effect. (The chemical basis for oxidation and the chemistry of paint were only partially understood during the nineteenth century.)

Zinc dust came into general use as a protective pigment in the early nineteenth century. Today 'zinc rich' coatings are considered amongst the best protective treatments available.

Linseed oil was a common binding agent of these protective pigments.

Pitch and *bitumen* had no anti-corrosive chemical properties but were used widely to form a reasonably dependable waterproof coating on structures such as bridges.

The *intermediate coats or undercoats* were applied to build up thickness and make it difficult for moisture or air to pass through the coatings. *The finish coat* of paint provided the first line of defence against the environment and determined the final appearance of the iron. Today the same principles of painting iron apply. The various coats within a painting system must, of course, be compatible with one another.

In interior locations, ironwork was often protected simply by the application of oils or waxes. A traditional treatment for wrought iron was to scrape, chip, or pickle the surface until all scale and foreign substances were removed. A heavy coat of linseed oil was applied, then the iron was heated, and wiped over with emery cloth. Finally a combination of beeswax and boiled linseed oil was rubbed into the surface. Items treated in this way were sometimes used externally, with the final preparation described being applied annually. Another annual coating sometimes given was varnish or shellac. Goose fat with lamp black is another treatment known to have been applied to wrought iron door furniture.

31

Paint analysis

The analysis of paint layers on ironwork should be part of a preparation and repainting programme. The analysis should be approached in a systematic way and undertaken by a person with experience in this field. Samples for microscopic analysis should be carefully removed by scalpel from protected areas which are representative of all coatings which have been applied. Where a less detailed analysis is appropriate, paint layers can be removed in sequence by rubbing with fine wet and dry carborundum paper or careful applications of a paint stripper such as methylene chloride in thixotropic form.

The need for good surface preparation for painting

Correct surface preparation is probably the most important single factor in the success or failure of a painting operation. Even the best paints or coatings may fail on a badly prepared surface, whilst the simplest and cheapest paint may perform well on a correctly prepared surface. Good surface preparation is essential for good adhesion. To achieve this the new paint must wet the prepared surface, and be applied to a firm, stable foundation which is free of contaminants such as grease and water soluble salts.

Specific contaminants to be removed

The preparation of a sound surface usually involves removal of old paint, rust, loose mill scale and soluble corrosion salts. It must be remembered that paint removal may reveal cracks, corrosion and casting defects which were not previously visible. Allowance should therefore be made at the outset for dealing with these.

Old paint and repainting

All paint which is loose, perished or flaking must be removed. It is not normally necessary to remove all previous paint coatings if these are sound, hard and firmly adherent and are known to be conventional drying oil paint types, unless a sophisticated modern paint is to be applied, e.g. two-pack epoxide resins. Sound paint surfaces may simply be rubbed down and refinished with one or two suitable coats. Only wet hand processes should be used because of the risk from dust from lead pigments. Rubbing down should remove residual gloss, surface deposits and blemishes (lumps).

Chipped areas of paintwork can similarly be rubbed down, ensuring that the surface under the paint to which corrosion has spread is also cleaned. Locally damaged paint areas should receive a shallow feathered edge, but it can be difficult to achieve a visually acceptable repair by this method to chips in surfaces which have received several layers of paint. New paint coatings should overlap at least 50 mm onto existing sound surrounding paint coatings, and must be compatible with the existing coatings.

Small areas of paint can be removed with thixotropic paint strippers such as methylene chloride. Their residues must be removed by white spirit or water, as appropriate. Flame cleaning and hot air blowers are also effective paint removers.

They must be used with care on thin cast iron because of the thermal stresses which can be set up by localized overheating.

Mill scale (wrought iron and steel)

Mill scale is formed as the result of the hot rolling of wrought iron and steel. As the sections leave the mill rolls they cool and the surface oxidizes, producing mill scale. Mill scale is a non-metallic, brittle surface which is easily damaged and tends to detach from the underlying metal. Rust can form at the break in the scale and spread sideways between it and the metal.

Loose or defective mill scale must be removed. However, there is evidence to suggest that wrought iron receives corrosion protection from sound, adherent mill scale. For this reason flame cleaning is the preferred treatment for cleaning wrought iron, as it will remove only loose mill scale.

Rust

Rust is an unsatisfactory base for paint and must be removed before protective coatings are applied. Rust which remains provides a source of further corrosion beneath new paint surfaces. Very small amounts of rust can be treated with a rust converter. However, thorough cleaning is always preferable and is recommended.

Soluble corrosion salts

Ferrous sulphate and ferrous chloride are undesirable water soluble salts which must be removed from the bottom of pits within an iron surface. They are not readily removed by cleaning with large-sized abrasive particles. Tests for soluble corrosion salts should be carried out on iron structures in marine and industrial environments both after cleaning and immediately prior to painting. A simple and effective method of testing for ferrous salts is the use of blotting paper dipped in a 10% solution of potassium ferricyanide (see BS 5493). The dried strips are applied to a dampened surface and will change colour in the presence of soluble ferrous salts.

Methods of preparing iron surfaces

Degreasing

Any oil or grease should be removed to avoid subsequent preparation methods spreading the contamination over a wider surface. Large quantities should be physically removed by scraping. The rest is best removed by warm water and detergent followed by thorough water rinsing. Non-caustic degreasing agents are also available, although a wipe over with white spirit and a succession of clean swabs often suffices.

Manual preparation

The simplest form of surface preparation of iron involves chipping, scraping and brushing with hand-held implements. While a surface prepared in this way may appear burnished and clean, only about 30% removal of rust and scale may be achieved. Scoring of valuable surfaces and loss of detail may also occur. Manual preparation may be useful at times in external situations as the first stage in inspection of a heavily corroded item, where alternative methods are not available. A corrosion-inhibiting primer such as red lead or zinc phosphate should then be used.

Mechanical preparation

These processes involve use of power-driven tools such as grinders and rotary wire brushes. A marginal improvement in efficiency over manual preparation can be achieved. Rust or other deposits in pits and crevices are rarely removed.

Needle guns, however, can be used successfully to remove rust and scale, which are broken up and loosened by the impact of a head of iron needles. They can reach into awkward corners and angles inaccessible to other equipment.

Flame cleaning

An oxyacetylene or oxypropane flame is passed across the iron. Both rust and loose mill scale quickly detach from the iron as the result of differential thermal movement. Immediately after the passage of the flame any loose mill scale, rust and dust that remains is removed by wire brushing.

Flame cleaning provides a high level of cleanliness from paint, loose mill scale and rust. It is the most appropriate method of cleaning for wrought iron. On a larger iron structure it can be a relatively slow method, however. On the other hand this method enables wrought structures to be examined carefully during cleaning for deterioration, missing elements and inappropriate past repairs. Furthermore, it is very effective at removing loose scale and rust from localized areas such as water traps, behind scrolls, water, leaves, etc.

Thin sections of wrought iron of less then 2 mm may warp during flame cleaning unless the method is used with care. Flame cleaning can be undertaken on site as the equipment is mobile. It can be used under relatively wet and damp conditions and helps to dry the surface. The method is, however, a fire hazard and if the flame is traversed too slowly, unbonded scale and other foreign matter may be fused to the surface.

Flame cleaning is often used to 'flash clean' an iron surface of any corrosion which may have developed in the time following cleaning and prior to the application of the primer. It can also be used to create a dry surface for the primer on small sections.

Acid pickling

Items are immersed in a bath of a suitable acid which dissolves or removes mill scale and rust. The acid does not appreciably attack the exposed surface unless the iron remains in the bath for an excessive time. On removal from the bath the iron must be thoroughly rinsed with clean water.

Warm dilute sulphuric acid or dilute phosphoric acid are normally used. Phosphoric acid has the additional advantage that the reaction with the iron results in a protective layer of phosphates on the surface (anodic inhibitors). Hydrochloric acid and sodium hydroxide (caustic soda) should not be used as they leave soluble salts on the metallic surface which promote corrosion at a later time.

Acid pickling is essentially a works process because it must be carefully controlled. Site application of acid washes is not recommended. The bath dimensions will determine the size of pieces which can be treated by this method.

Dry abrasive cleaning

Abrasive cleaning is an appropriate method for cleaning cast iron. It is also widely used on wrought iron because of the speed and relative cheapness of the process (it is much faster than flame cleaning). However, due to the softness of wrought iron it must be acknowledged that the milled or beaten surface will be removed or roughened if the approach is too heavy-handed. In conservation terms this can be undesirable and an alternative cleaning method should be considered.

In this method, airborne abrasive particles impinge on the surface, producing a roughened, cleaned surface. As with stone and brick cleaning, the success of abrasive cleaning of cast iron and wrought iron is highly dependent on careful work by skilled operatives, the right grits and the right supply of air pressure.

Abrasive cleaning of wrought iron requires a different approach from the cleaning of structural steel. The material is softer and needs to be cleaned at a slower pace. Test areas should always be carried out to determine the correct air pressure and size of grit. It is advisable to start at a pressure of 40 psi (6 kPa) with a fine grit, usually copper slag. A satisfactory cleaning pressure is not likely to exceed 60–70 psi (8–10 kPA).

The use of abrasive cleaning is usually not appropriate on wrought iron which includes fine details such as leaves and scrolls. The presure will need to be substantially reduced and a smaller nozzle used to ensure these are not deformed. Abrasive cleaning of wrought iron whose surface has high intrinsic value is not recommended as the abrasive particles will tend to 'peen' or rework this. Dry abrasive cleaning produces dust and debris on the surface of the iron which must be removed, preferably by vacuum cleaning, prior to painting.

Abrasive-cleaned surfaces are usually specified in terms of surface cleanliness and surface roughness. BS 4232, *Surface Finish of Blast Cleaned Steel for Painting*, and Swedish Standard SIS 055900, *Pictorial Surface Preparation Standards for Painting Steel Surfaces*, should be carefully interpreted before being applied to cast iron and wrought iron.

Wet abrasive cleaning

Wet abrasive cleaning is preferable to dry especially where lead-based paint is to be removed, as the dust problem is avoided. It is also useful in washing from the surface soluble iron salts such as chlorides and sulphates that form within deep corrosion pits. Wet abrasive cleaning is particularly suitable for the cleaning of iron structures in marine and heavily polluted environments. Cleaning should be done with a nozzle which has independent control over air, water and abrasive. This is essential for quick removal of slurry and good visibility of the work surface.

This facility is also important for the cleaning out of crevices and ledges and for the removal of excess water on completion of cleaning.

Wet abrasive cleaning may, however, cause unwanted water penetration at junctions. A water tolerant primer is usually advisable after washing or wet abrasive cleaning.

Precautions for wet and dry abrasive cleaning
For both methods care must be taken to mask surrounding surfaces. All caulking which is dislodged must be replaced. It is necessary to ensure that operatives are adequately protected and the potential environmental hazards such as dust, spent abrasive, and abrasive-laden run-off are dealt with properly.

Re-rusting of cleaned surfaces

Cast iron or wrought iron members which have been cleaned by flame or dry abrasive should be primed before rust starts to form. If this is not possible the surface should be flash cleaned immediately prior to priming.

After a wet abrasive cleaning method an iron surface will re-rust relatively quickly. Also, the surface may remain wet. It is possible to include a rust inhibitor in the final wash and hence delay the need to prime for up to 24 hours. The amount of inhibition must be carefully controlled (usually not greater than 5000 ppm) because the excess will cause the deposition of salts which will in turn cause the paint to peel. The water with inhibitor must be carefully removed from horizontal surfaces and water traps.

Rust removing solutions

Orthophosphoric acid is the basis of many rust remover solutions sold in retail outlets. Solutions which are described as 'chemically neutral' are based on a combination of acid and alkali materials and are available from suppliers of conservation materials. Best results are usually achieved when the corroded item is immersed, as this enables the treatment to break the bond between the base metal and the corrosion layer fully. Several solutions are available in gel form.

The mechanical removal of all loose and thick rust layers is advocated before the solutions are applied whenever this is archaeologically acceptable.

The importance of good site supervision

Proper site supervision by competent staff is important at all stages of work on a historical iron structure but in particular during the preparation for and application of paint. Test areas on all types of surface present, e.g. bars and leafwork, should be observed to ensure the right method or methods of cleaning are chosen. The details of each method should also be resolved at this stage.

2.8   PAINTING IRON

Paints for iron

A wide variety of paint systems suitable for painting cast iron and wrought iron are currently available, far beyond the scope of the traditional oil-based paints system. It is, therefore, possible to provide superior protection to iron, especially necessary in difficult environments.

Primers and inhibiting pigments

Red lead is still considered one of the best inhibiting pigments and is still available. If applied with suitable precautions its toxicity can be kept to an acceptable level (see BS 2523: 1966). Zinc phosphate is a more recent and more widely used inhibiting pigment which is non-toxic and has excellent inhibitive properties. Zinc dust depends for its efficacy on being directly in contact with clean metal to which it will give sacrificial protection if the substrate is subsequently exposed. A zinc dust primer may require a sealing coat and subsequent coatings need to be non-saponifiable, such as epoxies, chlorinated rubbers or vinyl paints.

As it is almost impossible to produce with one coat a continuous film of adequate and even thickness and free from pinholes (the points at which corrosion often begins), in all but mild conditions it is usually recommended that two coats of conventional primer (or a high performance primer) are applied. A second coat of primer rather than an extra finish coat can result in a longer life for the whole paint system.

Binders

The paint binder is the basic film-forming material which largely determines the range of uses of the resulting protective coating. The properties of binders available today can be summarized as follows:

Binders containing drying oils (Group F, BS 5493)
Drying oils such as the traditional raw, boiled or heat-treated linseed oils dry (harden) slowly by oxidation in the air, aided by drying agents. Oxidation continues slowly throughout the life of the paint until breakdown eventually occurs.

The following members of this group are listed in order of increasing water resistance and rate of drying: oleo-resinous, drying oil alkyds, urethane oils and epoxy esters.

Alkyd resin paints (drying oil alkyds) are probably the most commonly used paints. They will give satisfactory results in protecting internal or external ironwork if they are applied in sufficient thickness to a rust and scale-free surface, when the conditions to be withstood are not chemically corrosive or if no solvent resistance is required. Drying oil paints must not be used in continuously damp or alkaline environments, since their drying oil components will soften and dissolve by saponification. The performance of the drying oil systems can be improved by using micaceous iron oxide (MIO) pigmentation in undercoats and finishes (not in the primer — a compatible quick drying red lead or zinc phosphate should be used). The MIO does not provide corrosion inhibition, but its flakes produce additional protection by impeding moisture penetration and sheltering the binding medium from ultra-violet degradation. MIO paints were first introduced in the 1940s. They are a cost-effective paint but are only available in a very limited range of colours and not in white. Their surface finish is not a traditional one and may need over-coating for appearance's sake.

Single-pack chemical resistant binders (Group H, BS 5493)
These paints comprise polymers such as chlorinated rubber and vinyl in solvent. They dry by evaporation of the solvent (no oxidation). Chlorinated rubber paints are used where good resistance to water, acids and alkalis is required (but not to fats or a hot climate). Neither chlorinated rubber paints nor vinyl paints have the good surface-wetting properties of the oil-type binders and hence they require a high standard of surface preparation.

Two-pack chemical resistant binders (Group K, BS 5493)
In this category two or more components are mixed just before application. The chemical resistant binder is formed in situ as the result of a chemical reaction which is time- and temperature-dependent. For the protection of iron in corrosive environments, two-pack epoxy and polyurethane coatings and modifications of these with coal for an asphaltic pitch are most commonly used.

Metal coating of iron
Both zinc and aluminium coatings will provide cathodic protection to iron. Both metals can be applied by either hot dipping or metal spraying. They are a high-performance protection system which is relatively cost-effective. The surface must be clear of all dirt and mill scale.

Methods of applying paint
Brush application is the traditional method of applying paint to iron. It provides good contact between the liquid paint and the substrate, which results in good adhesion of the dry paint film. Dirt, dust or debris on an inadequately prepared surface will be incorporated into a paint layer during brushing. Painting by brush is preferred for coating rough, pitted or moulded surfaces to ensure penetration of paint into pits and other recesses. Lead-based paint should always be applied by brush. One-pack and two-pack chemical resistant binders are not always suited to brush application. However, brush application of other types of paint is often recommended for priming coats. While spray application of paints achieves a higher speed of work, the procedure requires a well-prepared surface and skilled operatives. Higher film thickness can be achieved by this method. Roller application should not be used for the initial priming coat of any paint system. It is suitable only for large flat surfaces.

Selection of a protective paint system

It is always advisable to consult a reputable paint manufacturer regarding the selection of a paint system, and the compatability of new and old paints. Selection of a protective system should consider the historic and archaeological value of the existing coating, the method of surface preparation, the environment of the structure, the proposed period between maintenance, difficulties of maintenance such as access, the conditions in which the paint will be applied and the availability of skilled operatives. It is always worth considering the upgrading of a paint system by the addition of an extra coat rather than the use of a more expensive paint, as similar results can often be achieved.

For good protection a paint system should provide a total film thickness of 125 microns to 250 microns. This normally means at least four coats of air drying paint. In all but mild conditions it is sensible to provide two coats of primer.

Conditions for applying paint to iron

Paints should never be applied to damp surfaces, unless they are specifically designed to do so. Condensation on all surfaces should be dried. Paints should not be applied during periods of rain, fog, snow or mist. In the UK, the months of November to February are generally not suitable for the outdoor painting of iron. It is, nevertheless, possible to select a coating system which is tolerant of these conditions.

The most frequent cause of bad paint adhesion is application to a damp, wet or frosty surface. Also, exposure of wet paint film to frost, rain, fog or dew before it has dried properly has a damaging effect on its performance.

Cast iron, wrought iron and fire

Cast iron elements are known to perform particularly well in fires. Many fires have been observed where the cast iron elements have survived intact while nearby wrought iron and steel has twisted or drooped and failed structurally. The ability of cast iron to withstand the onslaught of a fire is not consistently recognized by all local authorities at present. Damage can occur through quenching by fire hoses. Intumescent coatings can be applied to give cast iron an increased level of fire protection which to some authorities will be acceptable. While the intumescent coatings may substantially alter the appearance of ironwork they may be the only acceptable solution to the problem of fire protection. They provide a layer that under normal operating temperatures is only several millimetres thick but which, when subjected to a sudden severe increase in temperature, will rapidly expand and foam up to provide an incombustible layer of carbon up to 75 mm thick, providing the structural element with fire protection for the required period. (Intumescent coatings are often called intumescent mastics and should not be confused with intumescent paints, which will only provide spread of flame control of combustible substrates.)

Several types of intumescent coating exist within the constraints of fire protection required; principal criteria for the selection of a coating should include minimal thickness and appropriate texture. The coatings can be overpainted with most standard types of paint. They must be applied in a number of thin coats, otherwise they may slump. With straight and true profiles it is important to achieve an even finish.

The effectiveness of any intumescent coating is highly dependent on the preparation of the surface and the technique of application.

REFERENCES AND ORGANIZATIONS

References

1 Allen, Nicholas K C, 'The Care and Repair of Cast and Wrought Iron', Unpublished dissertation, Diploma in Conservation Studies, University of York, 1979.
2 Barraclough, K C, *Steelmaking Before Bessemer: Volume 1; Blister Steel — The Birth of an Industry*, The Metals Society, London, 1984.
3 Barraclough, K C, *Steelmaking Before Bessemer: Volume 2; Crucible Steel*, The Metals Society, London, 1984.
4 Barraclough, K C, *Bessemer and Sheffield Steelmaking*, Sheffield City Museums, Information Sheet 18.
5 Barraclough, K C, *Crucible Steel Manufacture*, Sheffield City Museums, Information Sheet 8.
6 Barraclough, K C, *The Origin of the British Steel Industry*, Sheffield City Museum, Information Sheet 7.

British Standards Institution:

7 BS 5493: 1977, *Code of Pratice for Protective Coating of Iron and Steel Structures against Corrosion*
8 BS 4232: 1967, *Surface Finish of Blast-Cleaned Steel For Painting*
9 BS 2569: Pt 1: 1964, *Protection of Iron and Steel by Aluminium and Zinc against Atmospheric Corrosion*
10 CP 3012: 1972, *Cleaning and Preparation of Metal Surfaces.*
Many paints for iron have their own British Standard: see BSI Index.

11 British Steel Corporation, *The Prevention of Corrosion on Structural Steels*, Technical Note, BSC Sections, Cleveland UK (no date).
12 Building Research Establishment, *Digest 70, Painting: Iron and Steel*, BRE, Watford, England, 1973.

Department of Industry and Institute of Corrosion Science and Technology *Guides to Practice in Corrosion Control:*

13 *12: Paint for the Protection of Structural Steelwork*, 1981.
14 *13: Surface Preparation for Painting*, 1982.
15 *14: Bimetallic Corrosion*, 1978.

16 Edwards, Ifor, *Davies Brothers, Gatesmiths — 18th Century Wrought Ironwork in Wales*, Welsh Arts Council/Crafts Advisory Committee, Cardiff, 1977.
17 Leigh, W J and Co, *Steel Protection in the 80's*, Notes from an illustrated talk available from W. J. Leigh & Co.
18 Lister, Raymond, *Decorative Cast Ironwork in Great Britain*, G Bell and Sons Limited, London, 1960.
19 Paint Research Association, *Quality Control Procedures when Blast Cleaning Steel*, PRA, Teddington, Middlesex, 1980.

20　Rivington's Series of Notes on Building Construction, *Notes on Building Construction, Part 3, Materials*, Chapter 4: Metals — Cast Iron, Wrought Iron and Steel, Longman, Green & Co, London, 1892.

21　Swedish Standard SIS 055900, *Swedish Standard Pictorial Surface Preparation Standards for Painting Steel Surfaces*.

22　Watkinson, F and Boniszewski, T, 'The Selection of Weld Metal for Wrought Iron', *BWRA Bulletin*, Vol 6, No. 9, September 1965.

See also the Technical Bibliography, Volume 5.

Organizations

British Cast Iron Research Association (BCIRA)
Alvechurch
Birmingham B48 7QB
Tel: (0507) 66414

British Foundry Association
Ridge House
Smallbrook Queensway
Birmingham B5 4JP
Tel: (021) 643 3377

The Welding Institute
Abington Hall
Abington
Cambridge CG1 6AL
Tel: (0223) 891162

Worshipful Company of Ironmongers
Ironmongers Hall
Barbican
London EC2Y 8AA
Tel: (01) 660 2725

Ironbridge Gorge Museum
Ironbridge
Telford
Shropshire TF8 7AW
Tel: (095245) 3522

Institute of Corrosion Science and Technology
Exeter House
48 Holloway Head
Birmingham B1 1NQ
Tel: (021) 622 1912

National Corrosion Service
National Physical Laboratory
Teddington
Middlesex TW11 0LW
Tel: (01) 977 3222

Paint Research Association
Waldegrave Road
Teddington
Middlesex TW11 8LD
Tel: (01) 977 4427

3 TRADITIONAL COPPER ROOFING

Sections 3.4 to 3.7 of this chapter were written by Dennis Toner, Senior Lecturer responsible for Plumbing Mechanical Engineering Services Section, South East London College, who was a consultant with the Copper Development Association from 1954 to 1976.

This chaper considers the utilitarian use of copper as roofing and flashing material on buildings. The initial cost of copper roofing was traditionally high but its length of service more than compensated for the price. As a result it was used in large areas only on major structures such as churches and public buildings. In small quantities it was widely used on more modest buildings for roof flashings and rainwater systems.

3.1 THE DEVELOPMENT OF COPPER MINING

Copper is the eighth most abundant metal in the earth's crust and is found in over 160 different minerals. Once the metal is extracted, copper is very durable and corrosion resistant.

England was naturally rich in copper and alloying metals, and their mining and smelting can be traced to prehistoric times. During the Bronze Age copper was used on its own (reference 6). As ores of tin and other metals often occur along with the copper minerals, the making of bronze and brass are thought to have begun with the accidental alloying of copper with tin and zinc respectively. While the oldest brass object is thought to date from 2000 BC, brass was certainly known in Egypt during the first century BC and in Rome from 20 BC–14 AD, when it was used in coinage (reference 6). The use of copper and bronze in the British Isles was well established before the Roman occupation. Crude copper implements found in Ireland are thought to date from 2500 BC, while bronze objects possibly as old as 1000 BC have been found in both England and Ireland (reference 5, p.23). The Romans mined copper extensively, exporting much of it, and gave rise to a considerable increase in the production of copper and bronze and development of the associated metallurgical arts. After the departure of the Romans the main cuprous metal industry to make headway was that of bronze bell-founding. This was a prosperous industry from the thirteenth century and the experience obtained was applied to the manufacture of cannon at the end of the fourteenth century (reference 1).

Following the discovery of rich copper deposits in Britain in 1566, the Mines Royal Act was passed in 1568. This dampened exploitation of England's copper resources by limiting mining rights to agents of the Crown. It was repealed in 1689, providing further impetus to the industry which led to the establishment of many metal factories. By the early eighteenth century Bristol and Birmingham had become centres of brass making. Copper mining was an important activity in many parts of the British Isles, the most important area being south-west England. In Cornwall copper mining and smelting began in prehistoric times. Between 1700 and 1850 it was of equal or greater importance than tin (reference 1, p.15). By the early nineteenth century Britain was the largest producer of copper in the world, providing 75% of the world's supply of copper (reference 10).

During this time the copper reverberatory furnace developed at Swansea in Wales became the world's largest producer (reference 1). By about 1840 copper mining in Great Britain had started to decline as the large deposits in Chile began to be developed, followed at intervals by Australia, North America, Spain and New Guinea.

3.2 PRODUCTION METHODS OF CUPROUS METALS

The traditional method of producing copper or brass sheet was the battery process. The metal was first cast into moulds. The ingots so produced were annealed before being beaten into sheets by means of hammers weighing up to 500lb. The sheets were then cut by water-powered shears.

During the 1700s the copper industry was mechanized. At the beginning of the century rolling machines began to supersede the battery system. Major advances were made in the drawing process which enabled complicated shapes such as architraves to be made. In 1769 a machine was invented which stamped patterns and designs into sheets of copper and brass. In 1839 the process of electroplating and electrotyping was invented (reference 10, p.46). The most recent major development was the invention of the extrusion process at the beginning of the twentieth century.

Tables 3.1 and 3.2 provide a useful summary of the relationship between copper, bronze and brass and the various methods of manufacture.

Table 3.1
Methods of manufacture — articles made of copper, bronze or brass

Repoussé	Copper	—	Brass
Cast	—	Bronze	Brass
Drawn	Copper	—	Brass
Stamped	Copper	—	Brass
Plated	—	—	Brass
Extruded	—	—	Brass

(Reference 10, p. 67)

Table 3.2
Methods of manufacture of cuprous articles — dates of use

Lost wax casting	Before 1600, after 1880
Sand casting	After 1540
Stamping	After 1761
Plating	After 1830
Extrusion	After 1900

(Reference 10, p.10)

In addition to these developments in fabrication methods, the copper alloy industry was greatly affected by upheavals in the science of metallurgy in the late eighteenth and nineteenth centuries. The William Champion patent of 1738 for making zinc from calamine ore meant melted zinc and copper metals could be mixed together for the first time in strictly controllable quantities. This led to the creation of many new alloys with special qualities to suit specific purposes, e.g. phosphor bronze (1871), silicon bronze (1882) and Monel metal (1905).

3.3   COPPER AS SHEET ROOFING

Light gauge copper sheet has been used for roofing and cladding for over 2000 years. The Parthenon was originally roofed in copper shingles, and copper roofing and cladding was used on several temples in India during this time. Today there are in existence, in Germany and in Scandinavia, copper roofs which were laid in the twelfth and thirteenth centuries with sheets, approximately 1 metre (3 feet) square, joined together with standing seams and double lock cross welts. There are also numerous examples of copper roofs throughout Europe which were completed after rolled sheet became available, and many are in top condition. A further example can be seen on the Chapel of St James's Palace, London, where the copper was laid in sheets approximately 1.8 metres×750 mm (6ft×2ft 6ins) and joined together with standing seams and double lock cross welts, the technique which is still known as the traditional method (reference 6, p.1).

The principal joint systems used to fix traditional copper roofing are illustrated in Figure 3.1.

The standing seam system is the truly traditional system of fixing copper sheet, although the term 'standing seam' is at times used to include the batten roll system. These terms refer to the type of joint between adjacent sheets of copper, in the direction of the slope. Generally the transverse joints were made by 'drips' or cross welts.

Up to the 1870s nearly all copper sheet roofing was fixed by the standing seam method. Only once it became commercially possible to produce thin sheets of consistent thickness was the introduction of wooden cores or 'batten rolls' into the fixing detail possible. The conical roll was preferred initially, and remained popular up until the 1930s. However, the conical roll system gave rise to difficulties at junctions, particularly at abutments and drips, which led to cracking.

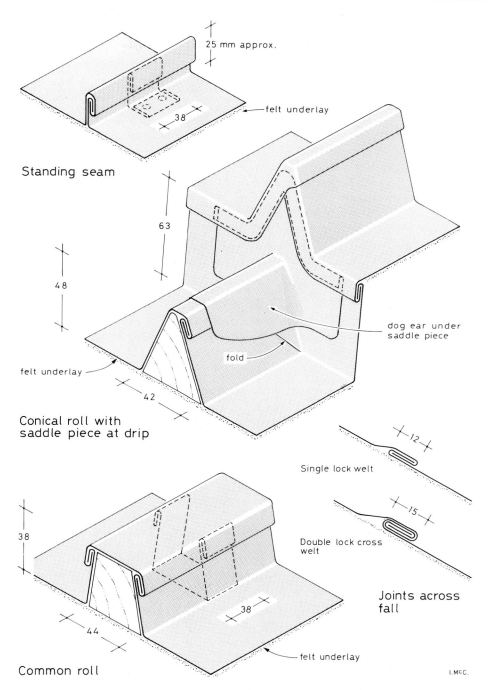

25 mm approx.

felt underlay

38

Standing seam

63

48

felt underlay

42

Conical roll with
saddle piece at drip

dog ear under
saddle piece

fold

Single lock welt

12

Double lock cross
welt

15

Joints across
fall

38

38

44

Common roll

felt underlay

I.McC.

**Figure 3.1   The principal joint systems used to fix traditional copper
roofing.**

Many such problems became evident around 1950 and the batten roll, with its near-square wood batten and copper capping strip, became the predominant wood core fixing system. The batten roll had begun to be used in the late nineteenth century and it can still be seen on the roofs of buildings such as seafront kiosks, shelters and piers.

In the traditional methods of fixing copper, silver brazing or soft solder were used for small joints such as within weathering details but never for repairs or jointing between roof sheets where there was any likelihood of stress being imposed on the join. These principles apply in good traditional copper practice today.

The relatively small and irregular areas of traditional buildings were suited to the traditional detailing systems of copper roofing. In the early 1900s interest began to be shown in using larger and thinner pieces of copper and the era of the development of new copper roofing systems began. The most common of these is the Long Strip System, which began to be used in the United Kingdom in 1957, although prior to that it had been practised on the Continent, particularly in Switzerland, over a period of many years. The system, comprised of preformed panels 520 mm (1ft 10½ ins) wide up to 8.5 metres (28 feet) long, is particularly suited to large areas of plain roof.

The principles of good copperwork in new roofing of the traditional and long strip methods are clearly presented in *Copper in Roofing – Design and Installation*, Technical Note 32, issued by the Copper Development Association. Here we deal only with the repair and maintenance of existing, historical traditional copper roofing.

3.4   THE BEHAVIOUR OF COPPER

Copper is one of the most corrosion-resistant architectural metals. It is nevertheless attacked by alkalis, such as ammonia and various sulphurous compounds, that can combine with water to form sulphuric acid.

Copper is not usually subject to galvanic corrosion. However, other metals in contact with copper such as iron, steel, zinc, aluminium, or galvanized steel may well become corroded. Direct contact between copper and other metals is best avoided. For aesthetic reasons steel should not be mounted above copper components because of the disfiguring stains which will develop on the copper.

Copper can be affected by acidic run-off from cedar shingles and lichen. This aspect is discussed in depth below, with appropriate remedial actions.

Copper can also be affected by acidic water run-off from bituminous surfaces, but this is usually only a problem where a small area of copper, such as a gutter, receives run-off from a large area of bituminous roofing. The repair and maintenance of copper roofing can, at times, require the use of a small area of bituminous paint, but this would only be above a large area of copper roofing. Urban rainwater, already considerably acidic, will be only minutely affected during its contact with the bitumen. The benefits of this occasional and restricted use of bitumen far outweigh the disadvantages.

Copper weathering and patina

The development of patina on copper depends upon the impurities in the atmosphere, and these vary with each locality. Even in the same locality the weathering pattern can vary according to the degree of exposure of the copper to wind and rain. If the copper is left to weather in an unpolluted environment it will first become coppery red, gradually deepening to a dark brown. In a lightly corrosive atmosphere copper acquires a patina which is predominantly green copper carbonate. As this material has similar expansion properties to the copper it adheres to it and provides the metal with protection from further corrosion. In places near the sea a pale green layer of basic copper chloride can be expected.

However, in a highly polluted environment, copper can acquire a colour ranging from various shades of green through to blue, brown or black. These are the colours of the copper corrosion products formed by the electrochemical reaction between the copper and the several severe agents in an urban/industrial environment.

Apart from the different colours produced, the ultimate effect on the copper may also be different. In some cases the corrosion is progressive, while in others a firmly adherent protective film is formed. In an urban or industrial environment the corrosion products often encourage further corrosion. When the products of corrosion are water-soluble a fresh surface of copper is constantly being exposed to attack.

Verdigris

The term 'verdigris' should not be used interchangeably with the term 'patina'. The terms refer to the products of two different reactions. Verdigris is caused only by the chemical reaction of copper with acetic acid and is composed of a mixture of basic copper (II) acetates. In contrast to the copper salts of natural patina, verdigris is water soluble. Visually verdigris may be recognized by its striking green colour (reference 4, p.55).

3.5   DEFECTS AND DAMAGE*

Provided a copper roof is correctly designed and laid, a long trouble-free life invariably ensues. Where design or workmanship does not meet required standards, problems can arise ranging from localized leakage to total roof failure. Occasionally storm damage or accidental damage by falling objects such as masonry will occur, and even a trouble-free roof will eventually require maintenance or repair work of one form or another in order to extend its useful life.

Copper roofing is highly specialized and, unless faults and defects are correctly diagnosed and dealt with, further damage may be inflicted by misguided repairs. Therefore it is vitally important that architects and others responsible for the maintenance and repair of damaged copper roofs should be able to recognize the various types of defects and recommend appropriate remedial treatment.

* D. E. Toner is the author of the text and figures from this point up to and including section 3.7, and retains the copyright therein.

3.6 INSPECTION

Outward signs of deterioration and damage to a copper roof do not always indicate the full extent of problems beneath the covering. It is advisable to arrange for a copper roofing specialist to be in attendance at an inspection to open seams and welts as required, and to close them again upon completion of the inspection.

Suitable means of access to a roof must always be provided, but it is usually quite safe to walk on flat or low pitch roofs (up to 20 degrees), provided reasonable care is exercised and the surface is dry. The surface of wet copper is extremely slippery, and whenever possible inspections should be delayed until conditions improve. Rubber or plastic soled shoes are more likely to slip on wet copper than rope or leather soles.

It is advisable to include a wire gauge or micrometer in the inspection kit for checking thickness of sheet, and a moisture meter for testing dampness levels of walls and roof structure. A pair of binoculars is particularly useful for viewing inaccessible parts of a roof (spires, fleches, etc.) and a mirror and torch will prove invaluable for dark roof spaces and voids. A camera enables the situation to be recorded.

If available, a copy of the roof plan showing the layout of the copperwork will facilitate the accurate recording of the type and position of defects, relevant dimensions, etc. Should it be necessary to remove small samples of the sheeting for further tests, e.g. metallographic examination, temporary patching may be carried out using 75 to 100 mm (3–4in) wide adhesive weathering strip.

3.7 CAUSES OF FAILURE AND METHODS OF REPAIR

Mechanical damage, corrosion attack and deterioration of the understructure are the usual causes of premature failure and long-term deterioration of copper roofs and weatherings.

The extent to which each category may affect a covering can vary greatly, and more than one category may be involved. Repairs and remedial treatment will depend upon the precise nature and extent of the defects. The categories have been subdivided as follows. Appropriate repair techniques are discussed within each section.

Mechanical damage
- Wind displacement
- Excessive thermal movement
- Restriction of thermal movement
- Accidental damage

Corrosion attack
- Acidified rainwater run-off from other surfaces
- Concentrated flue gases

Deterioration of the understructure
- Wood-boring insects and rot
- Condensation

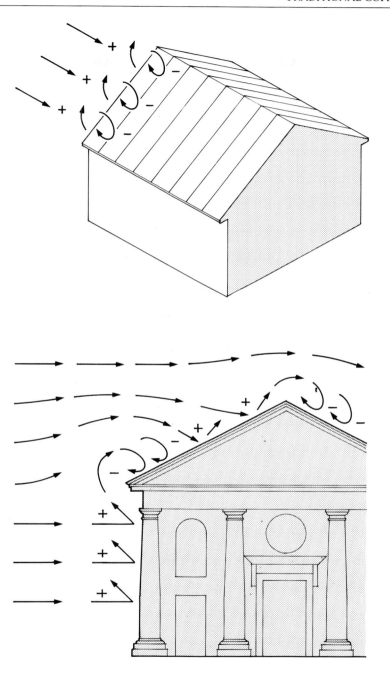

Figure 3.2   The negative wind loading on the eaves, ridges and verges of copper roofing

49

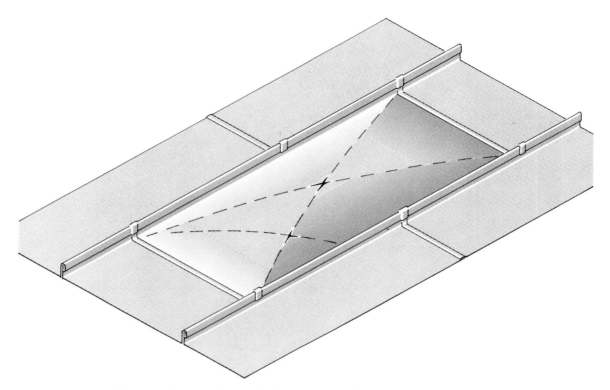

Figure 3.3 'Star' cracks in a copper bay

Mechanical damage

Wind displacement

This is the most common cause of mechanical deterioration of copper roofing. The loading imposed above a roof by a 100 mph (160 kph) wind gust of 3 seconds' duration can vary between $+88$ kgf/m^2 and -151 kgf/m^2 depending upon the direction of the wind, the topography of the surrounding area, and the height and angle of the roof (BS CP3: Chapter V: Part 2: 1972). At eaves, ridges and verges (Figure 3.2), the negative loading can rise to approximately 300 kgf/m^2 (CP3: Chapter V: Part 2: 1972).

The weight of most roof materials often makes a significant contribution to the restraining forces needed to resist such negative wind loading. However, in this respect the weight of 0.6 mm (24 SWG) and 0.7 mm (22 SWG) thick sheet copper used for the traditional standing seam and batten roll systems of copper roofing is comparatively low, and their weights of 5.4 kg/m^2 and 6.3 kg/m^2 respectively contribute little towards the stability of a copper roof. Similarly, the bays possess little natural strength or rigidity to resist windlift, since they are made of soft temper copper sheet to facilitate hand fabrication and are not profiled, fixed, or bonded in the areas between joints because of the unacceptable jointing and

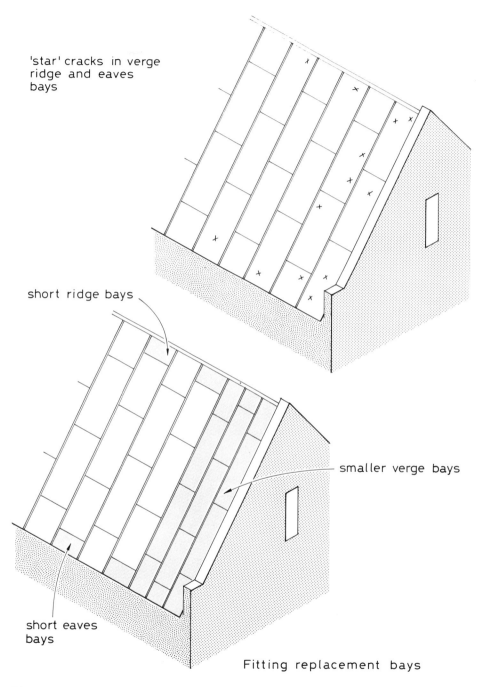

'star' cracks in verge ridge and eaves bays

short ridge bays

smaller verge bays

short eaves bays

Fitting replacement bays

Figure 3.4   The formation of 'star' cracks in bays at verges, ridges and eaves and the remedial solution of modified bay sizes

51

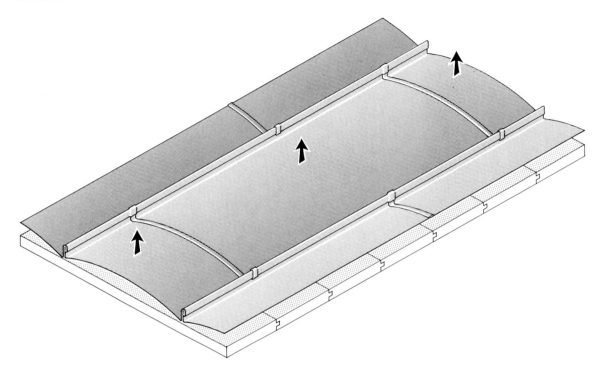

Figure 3.5   Wind lift of panels with no cross clips

thermal movement problems which would arise. In such circumstances bay size is of paramount importance in limiting vulnerable 'free' areas between joints. This limitation, together with strong fastenings in the joints and wind-tightness in the covering and decking, is essential if high wind loading forces are to be successfully countered. Many failures are directly attributable to a lack of understanding of these basic points.

(1) 'Star' cracks This term refers to the characteristic star shape of cracks produced by repeated flexing or bending of a bay sheet under wind load or thermal stress. The lines of fold in a rectangular sheet held around its periphery will tend to be diagonal. Where the lines cross, stresses are raised and it is at these intersections that 'star' cracks develop (Figure 3.3).

If left unchecked 'star' cracks will extend to become large splits allowing ingress of rain with all of its attendant problems, and creating a potentially far more serious condition by allowing wind to penetrate the covering. The combined wind loading above and below the sheets can lead to large areas of the covering becoming detached from the decking, and then torn from the roof by a particularly high gust.

Usually there is ample warning of the onset of such trouble, and provided prompt action is taken the more extreme forms of damage may be averted. An early symptom of wind-displacement trouble, often reported by the occupants of the building, is known as 'wind-drumming', that is, vibration or oscillation of the sheets producing noises similar to a distant rumble of thunder or loud drumming sounds as the sheets strike the roof deck.

Remedies for 'star' cracks Provided 'star' cracks are confined to localized areas such as the verge, ridge or eaves of a roof, then replacement of the damaged sheets with narrower and possibly shorter verge bays, and shorter ridge and eaves bays, will normally cure the problem (Figure 3.4). Replacement verge bays should not exceed a net width of 380 mm (1ft 3ins) with double lock cross welts at not more than 920 mm (3ft 0in) centres. Damaged ridge and eaves bays will require individual treatment suited to their specific condition. It is difficult to lay down hard and fast rules, but as a general guide the section containing the 'star' cracks should be removed or the original bay shortened by at least a third, whichever is the greater. If cracks are widely distributed in a bay it should be replaced with two new bays each approximately half the length of the original.

The worked edges of the existing sheets will be in a hardened condition and may have sustained some damage during the unfolding of welts and seams. Where possible these edges should be cut back to flat unworked metal before joining to the replacement bays. Any worked edges which must be retained should first be dressed flat and then annealed before attempting further working. As annealing takes place at temperatures well in excess of the combustion temperatures of timber, felt underlay and certain other roofing materials, all necessary fire precautions must be observed. In particular fire resistant insulation pads or sheeting should be inserted between the copper and combustible material during annealing and left in place until the copper has cooled. An efficient fire extinguisher must be on hand at all times, and no annealing should be carried out later than two hours before finishing work for the day.

Existing copper clips and continuous fixing strips should not be re-used in repair work. New clips and fixing strips of the recommended sizes (BSCP 143: Part 12: 1970 *Sheet and Wall Coverings, Copper*) should be made of the same thickness sheet copper as that used for the repair. Similarly, all nails and screws used to fasten the clips, fixing strips, wood rolls etc. should be new and of the approved type (BSCP 143: Part 12: 1970).

Any other contributory faults revealed during the inspection, or which may come to light whilst repair work is in progress, e.g. poorly designed roof terminations and 'gapped' boarding permitting wind penetration, or decayed decking, should be rectified at the same time.

If 'star' cracks are numerous and widely distributed over a roof, then the causes are likely to be of a general nature and piecemeal replacement of damaged bays will not be the answer, since further outbreaks of the same trouble could be expected in the remaining undamaged bays. Stripping and re-covering may be the most economic and reliable course in such cases.

(2) Wind lifting of bays and seams The component parts of a copper covering are firmly locked together by welts and seams which prevent easy separation of one

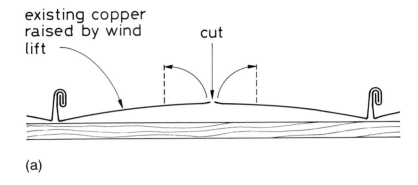

(a)

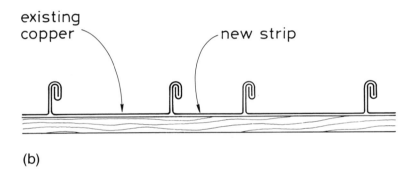

(b)

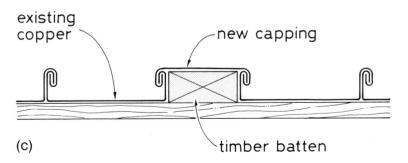

(c)

Figure 3.6   The refixing of bays affected by wind lift due to no cross clips

unit from another. It follows that, should any part of a copper roof be lifted from the decking by wind, repeated movement of that part can cause the problem to spread to adjoining bays, with varying consequences.

(3) Wind lifting of central zone of panel - no cross welt clips Where cross welt clips have been omitted, the central zones of entire panels can be pulled upwards, initially by as much as 25–30 mm (1in–1¼in) in the centre of 'bow' (Figure 3.5). If left, the central zone may be pulled up to 100–150 mm (4ins–6ins), at which stage upstands to standing seams and roll joints are pulled out and the sheets are held only at the extreme ends of the fixing clips. When this stage has been reached the sheets will be so badly damaged as to render them suitable only for scrap.

If the damage is intercepted at the initial 'bowed' stage, however, a number of options are open. In the case of a small area involving perhaps two or three bays in one panel, the cross welts may be carefully prised open, permitting a 50 mm (2ins) wide clip to be inserted and fastened to the understructure by copper nails or brass screws. Each welt which is opened should first be annealed whilst the copper is in the raised position, using damp rags on both sides of the bay to keep the metal cool where it is in contact or close proximity to the decking. The opened edges should be re-annealed before dressing the bay sheets down and closing the cross welt with the new clip incorporated.

More frequently, larger areas of roofs with unclipped cross welts are affected, and other methods must be employed to rectify the problem. Providing the damage has not progressed beyond the initial 'bowed' stage, repairs may be carried out as follows:

(a) Each raised panel is cut down the centre from ridge to eaves and the edges are turned back to form upstands to suit either a standing seam or a rectangular timber batten (Figure 3.6(a)).

(b) Where the timber batten method is used, the batten should be fastened with countersunk head steel screws into the decking. Fixing clips should pass under the batten and be turned up on each side and folded in with the capping strip welt similar to the method adopted with a square roll joint. A new capping strip will be required to cloak the top of each timber batten, and this should be formed from fully annealed copper strip of the same thickness as the existing roof covering. The timber batten method of repair is shown applied to a standing seam roof (Figure 3.6(b)), but is perhaps more suited to the batten roll system, where the repair blends in better and is less noticeable.

(c) With the standing seam method of repair a new narrow strip is prepared with undercloak and overcloak upstands to complement those turned on the existing sheets (Figure 3.6(c)). New clips are spaced at a maximum 380 mm (1ft 3ins) centres along the length of the new seams and fastened to the understructure by two copper nails or two brass screws per clip. Clips should not be placed at junctions with cross welts to avoid building up unmanageable thicknesses of copper in the standing seams. Provision must be made in the new strip for longitudinal expansion movement to take place. A convenient and visually acceptable way of accomplishing this is to incorporate 'dummy' welts in the strip similar to the method illustrated in Figure 3.14, Section A–A.

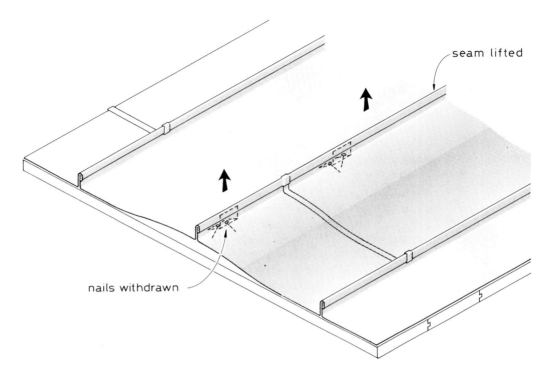

seam lifted

nails withdrawn

Figure 3.7   Wind lift of panels due to a detached seam

They should be arranged to match the staggered pattern of welts in the existing covering.

(4) Detached seam A detached standing seam (Figure 3.7) may require only minor remedial treatment or it could necessitate the complete stripping and recovering of the roof. The main points to be established are whether the fixings in the seam have failed due to withdrawal of the nails from the decking, whether nails have pulled through the holes in the clips, whether insufficient clips were provided in the seams, or whether breakage of the clips has taken place. Also, it will be necessary to establish whether the faults are localized or common to the whole roof covering. This will entail opening a number of seams and retrieving sample clips and nails for examination.

If the nails fastening the clips are found to be cut copper tacks, or copper nails with a small head, instead of special large flat head copper wire nails with a barbed shank as specified in the relevant section of CP 143: Part 12: 1970, then localized repairs are unlikely to prevent a recurrence of the trouble in other locations if similar conditions apply. Cut copper tacks are tapered throughout their length, and even slight withdrawal will seriously reduce their hold in the decking. Shrinkage of a timber roof deck during service due to natural drying out will inevitably exacerbate the situation.

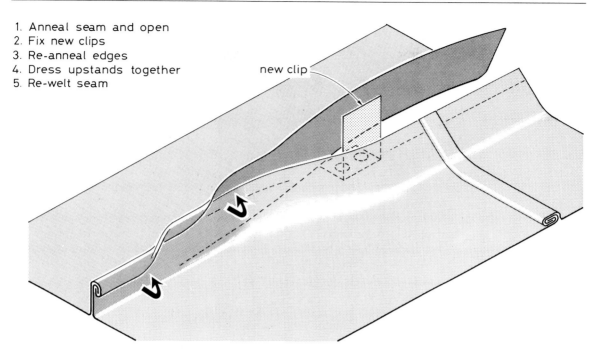

1. Anneal seam and open
2. Fix new clips
3. Re-anneal edges
4. Dress upstands together
5. Re-welt seam

new clip

Figure 3.8   Refixing a detached seam affected by wind lift

A copper nail with a small head is almost as unreliable in service as a tack. As a nail is driven through a clip into the decking it first forces an indentation in the clip before the resistance of the underfelt and timber is sufficient for the nail to pierce the metal. Where large head nails are used, the effectiveness of the fastening is not significantly impaired. By contrast, a nail with a small head offers minimal resistance to being pulled through the tapered hole in a copper clip. Where nails or tacks are at fault, any savings which might be achieved by carrying out limited repairs, compared with more drastic and costly treatment, will have to be carefully weighed against the risk of further trouble and subsequent damage to the building and its contents.

If a seam is insufficiently clipped, and some of the nails in existing clips coincide with joints or splits in boards, and there is localized high wind turbulence, individual standing seams can become detached. The adjacent seams and welts may well be firmly fixed and the surrounding areas of copper undamaged, in which event remedial treatment similar to the refixing of the centre of a panel (Figure 3.6) could be applied where the whole length of seam from ridge to eaves has to be dealt with. In the case of a relatively short length of deatached seam (2 to 3 m, 6ft 6ins to 10ft 0ins approx.), refixing along the lines shown and described in Figure 3.8 may well be the most economic repair. It will be noted that the repair involves annealing the existing copper both before and after opening the seam. Accordingly, fire precautions as previously described must be strictly observed.

Where broken clips are found in a detached standing seam, the number and size of the clips and the method of fastening them should be checked. The minimum width for a standing seam clip is 38 mm (1½ins). The clips should be spaced at a maximum 380 mm (1ft 3ins) and be fastened to the understructure by two copper nails close to the turn up. There is a good safety margin in such provision, but where clips have been omitted or their width significantly reduced, e.g. down to 25 mm (1in), or they are nailed at the tail instead of close to the turn up, clips may break under the repeated strain of wind loading. Provided such deficiencies in clipping are not general to the whole copper covering the methods of repair shown in Figures 3.6 and 3.8 are again relevant. If the deficiencies are general, re-covering the roof may be the only satisfactory option, since the risk of serious damage to the copper sheeting by severe storms would be high, with a correspondingly high risk of injury to the public.

Clearly it is not feasible to open every seam or welt on a roof to establish how, or whether, it is fastened, but a number of easily carried out checks yield useful evidence which helps to build up a picture of the state of the fixing system. A detached seam where the nails in the fixing clip have withdrawn from the decking is clearly distinguishable from a seam suffering from a broken clip. The nails, being unable to fall out of the clip due to the sheeting above, turn on their side and prevent the seam either resting on or being pressed flat against the top surface of the decking. Where a detached seam is found to be deficient in the number of clips provided, the extent of the problem can often be determined by noting the positions of the clips in the opened seam, and then closely examining the exterior of the other seams at approximately the same intervals in their length. The additional thickness of the folded clip frequently produces a slight bulge in the seam which is not too difficult to see when the observer knows what to look for. Nails in the tails of clips instead of close to the turn up will allow a standing seam gripped by a pair of pliers to be raised from the decking without the application of undue force, and by varying amounts depending on the length of tail and the position of the nails. A similar lifting test performed on cross welts, but using a broad-blade screwdriver or a narrow floor bolster to hook under the open edge of the welt, will reveal whether it contains a fixing clip. In addition, the outline of the clip is also usually discernible in the cross welt as with the standing seam.

(5) The detached batten roll − causes and repair Wind lift problems are encountered far less frequently with batten roll copper roofs than with standing seam roofs, for a number of reasons. A copper roof with batten roll joints is stronger in certain respects than its standing seam counterpart. It has a deeper profile than a standing seam, and this enhances the rigidity of the bays. More material is required in the upstands to the rolls, so the net width of bay for a given width of sheet or strip is less than with standing seams, and this again gives added strength to the system. On the rare occasion when a batten roll copper roof has been wind damaged, more often than not the trouble is found to be unclipped cross welts (considered above). If a batten roll has pulled away from the decking the cause is almost certain to be bad workmanship.

Standard practice is to secure wood rolls with No 12 gauge countersunk head steel screws of sufficient length to obtain a full 25 mm (1in) hold in the nominal 25 mm (1in) thick timber boarding, and for the screws to be spaced at 450 mm (1ft 6ins) centres along the rolls. If shorter or lighter gauge screws are used or their spacing is significantly increased, or if nails are used instead of screws, then a detached roll may result.

There are three very effective methods of securing a detached roll. The method selected depends to some extent upon the design of the roof, visual consideration and cost.

(a)   Where access is available to the roof space (usually roofs pitched at 35 degrees and above), refixing of the roll can be carried out from the underside of the boarding. It is first necessary to relocate the screws or nails in the original holes. This is not as difficult as it sounds since the rolls cannot move out of line in a longitudinal direction and rotation of the roll is minimal, so the points of nails and screws usually descend straight into the holes.

The rolls and fastenings must next be driven down until the underside of the roll is in close contact with the boarding. A hardwood block should be used on top of the roll to avoid damage to the capping, particular attention being paid to any fastenings which are forced back up through the roll, their presence being advertised by small 'bumps' which appear in the capping.

If the existing roll fastenings pierced the underside of the boarding, the line of the roll will be visible, and screwing up into the roll with new screws should be a straightforward matter. On the other hand, if the fixings were short, the centre of the roll at the top and bottom of the slope must be determined. A convenient way of accomplishing this is to drill a small hole (6 mm (¼ins) diameter maximum) at each position, passing through the centre of the capping and roll to the underside of the boarding. A chalk line struck between the two points, and pilot holes drilled through the boarding at the appropriate centres along the line, plus downward pressure exerted on the roll by an assistant on the upper surface,will assist in the speedy refixing of the roll. The two small holes left in the capping may be patched with a small disc of sheet copper about the size of a one penny coin, the facing surfaces having first been cleaned and pretinned with soft solder and a copper bit before they are finally 'sweated' together.

Should the roll prove to be located directly above a joist, the methods described in either (b) or (c) should be used.

(b)   With flat or low pitch roofs, refixing from the underside is seldom an option due to lack of access. Remedial work must of necessity be carried out from the top surface.

Removal of the capping strip by carefully prising open the single welts attaching it to the bay upstands on either side of the roll will expose the roll top and fastenings. After the wood-roll has been driven back into position, any of the existing fastenings forced upwards through the roll may be extracted and replaced with new, more substantial screws.

Alternatively, the existing fastenings can be driven back below the top surface of the roll, and additional screws provided in different positions.

The replacement capping strip should be prepared from new copper of the same thickness as the original. It is usually impossible to salvage the original capping strip, as it will have been buckled and hardened during removal, and annealing and straightening will destroy any green patina which is the main reason for its re-use. The half-welt edges of the bays should be dressed flat on the support of a metal bar before the new capping is welted in, but on no account should an attempt be made to anneal the edges, as the wood-roll will ignite.

(c)   Another method of refixing a detached wood-roll where access is confined to the top surface of the roof also has the advantage of minimal disturbance to the existing copper covering and hence any green patina which may have formed.

As with the previous two methods it is first necessary to restore the roll to its original position (flat against the boards) with the aid of a hardwood block and club hammer. Using a 12–16 mm (½–⅝in) diameter hole saw and electric drill, holes are cut in the copper capping at the required centres. Drilling is continued through the wood roll with a 6 mm (¼in) twist bit and the hole countersunk in the top of the roll. The roll may then be refastened using screws of appropriate gauge and length. The repair is completed by soldering copper discs 25–30 mm (1–1¼″) in diameter over the holes in the capping strip. The small discs are unlikely to create a strong visual contrast when seen from ground level, but if required they can be made to blend in with a well weathered copper roof by artificial patination (see Section 3.7).

Variations of the above methods are possible. For example, bolts could be used instead of screws, the bolts being fastened with a washer and two nuts on the underside of the board, with one of the nuts acting as a lock nut. Such a method would be particularly suited to roofs with low density insulation board deckings, always providing access is available to the underside of the boards. Long screws, driven right through the rolls and boards and into a 25 mm (1in) thick wood block measuring 100 mm by 100 mm (4in by 4in) held on the underside, are a further possiblility for insulation board deckings.

Inevitably, some of the existing fastenings will not re-enter the original holes in the boards and will offer considerable resistance to being 'hammered home'. Should this lead to perforation of the copper above the head of the nail or screw, a soldered copper disc repair will restore weathertightness to the capping.

(6) *Wind damage of the ridge and its repair* Reference has already been made to the increased wind loading which occurs at a ridge. For this reason it is good practice to use a ridge roll in preference to a standing seam ridge, because it adds strength to the covering where it is needed. In spite of this, standing seam ridges were sometimes used, generally for reasons of economy. Where ridge bays have suffered 'star' cracking due to wind displacement, the damage can extend to the ridge seam. This is a particular problem where the need for clips in the ridge seam has been overlooked.

The recommended method of repair is to cut out the entire run of standing seam ridge, undamaged as well as damaged sections, turn back the edges of the ridge bays

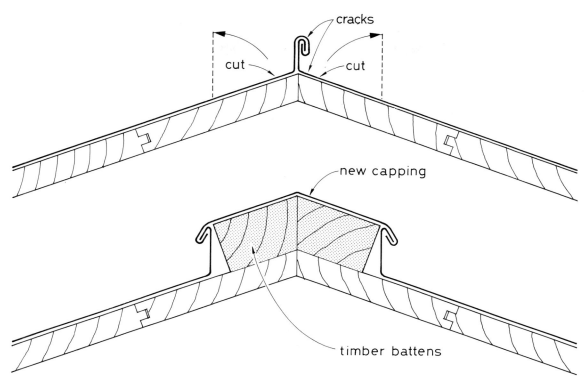

Figure 3.9   The repair of wind damage of a ridge

to form upstands to a new two-piece timber ridge roll, and welt on a new copper capping as shown in Figure 3.9. Fastenings for the timber battens should be long screws driven through the roof board into the ridge board or rafters. Clips along the ridge (not shown in Figure 3.9) should be 50 mm (2 in) wide and placed at two per bay. They may pass under the ridge roll and turn up on each side, or be nailed on the side of each section of the roll before fastening them.

Excessive thermal movement

As a result of the experience gained over very many years, well defined limits for the maximum sizes of copper sheets for traditional systems of roofing have been established. Provided these are not exceeded, thermal movement in the covering is seldom a problem since the small dimensional variations in bays and gutters are safely absorbed in the multitude of seams, welts and other joints which hold the sheets together. The one exception is where the standing seam system is used on long plain roof slopes. Unless special provision is made for thermal movement, cracks in the seams can result. Other thermal movement failures have come to light which would seem to indicate that some designers and installers were under the impression that the permissible length for a copper bay or gutter was limited only by the length of strip which could be obtained from the manufacturers!

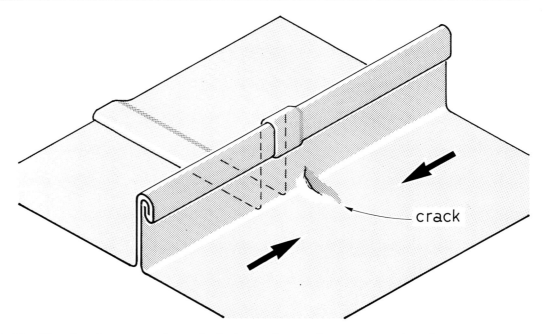

crack

Figure 3.10   The development of cracks in long slopes due to excessive thermal movement

(1) Long slopes The ability of a standing seam roof to absorb longitudinal movement, i.e. parallel with the slope, can only be relied on up to a maximum of about 9 metres (30 feet approx.). Above this length the cumulative movement of the seams appears to be too much for the system to accommodate on a long term basis. The alternating compression and tension stresses created by heating and cooling cycles cause cracks to develop in the seam on the opposite side to a cross welt. They are usually located in the angle of turn up, and at right angles to the standing seam (Figure 3.10). The defects are nearly always found to be concentrated in a central zone across the roof equal to about a third of the length of the slope. The problem appears not to affect batten roll systems of copper roofing, presumably because the wood-roll separates the upstands of adjoining bays, and all bays are free to move independently.

In the early stages of failure, small copper disc patches may be soft soldered over the cracks, gaining a useful reprieve of several years before the next batch of cracks appears (Figure 3.11). Eventually the cause must be eradicated if widespread failure is to be averted, and this will involve introducing one or more expansion joints across the slope to limit the maximum straight run of standing seam to 9 metres (30 feet approximately).

An economic and practical method of introducing a transverse expansion joint on an existing roof is shown in Figure 3.12. Two cuts are made across the slope and a section of covering is removed of sufficient width to allow a timber angle fillet to be fastened to the boards. The size of the fillet will depend on the pitch of the

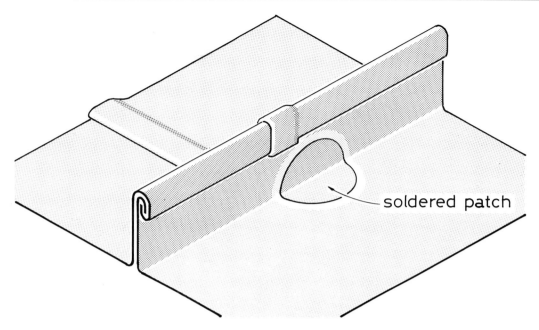

soldered patch

Figure 3.11   Excessive thermal movement — soldering a patch over a crack as an interim expedient

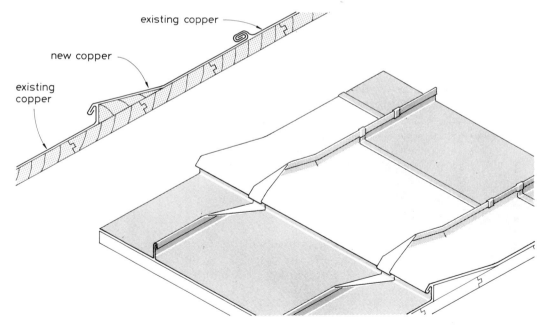

existing copper

new copper

existing copper

Figure 3.12   Excessive thermal movement — the insertion of a transverse expansion joint in a long slope

roof and in some cases it may have to be formed by two pieces of timber, but the vertical height at the front of the fillet should be a minimum 35 mm (1⅜ins), with a taper up the slope to give a good fall across the fillet. The ends of existing seams on the lower part of the covering should be turned over so that an upstand may be formed to go against the front of the angle fillet. The ends of the seams on the upper part of the covering are annealed for a distance of about 300 mm (1ft 0in) and carefully unfolded. Alternate bays are then cut back by about 150 mm (6ins) in order to create a stagger in the cross welts. After a further anneal, additional clips are fastened in position, new short bays joined to the existing sheets with a double lock cross welt, and the standing seams and drip-edge welt are completed as shown.

Scrupulous observance of fire precautions is once again necessary during all annealing operations.

(2) Long gutters The difficulty and cost of providing recommended falls, drips, cesspools and chute outlets in parapet and box gutters is a problem most designers will have encountered and few resolved according to the book on every occasion. Reason dictates that there must be a safety margin in the maximum lengths quoted in various authoritative documents and so, at times, small extensions of gutter length, as well as slight reduction of falls and the number of drips, are made, which enables significant savings to be made in the roof timbers. The real dilemma in such cases is in deciding where the line between safe economy and risky practice may be drawn without creating another similar problem.

A typical failure due to excessive thermal movement in a long gutter is shown in Figure 3.13. Referred to as a 'pinch' crack, it starts as a small wrinkle in the angle between the gutter sole and an upstand. Repeated expansion and contraction cycles enlarge the wrinkle and pinch the folds in the copper, causing the metal to harden severely and eventually crack. Dressing the wrinkled copper flat and soldering a copper patch over the crack will prove ineffectual in the long term because the trouble simply transfers elsewhere.

The Copper Development Association published data on the design of long copper gutter linings as long ago as 1947. This has been updated at regular intervals and is currently included in their latest booklet *Copper in Roofing – Design and Installation*.

Briefly the principle involves using quarter-hard temper copper instead of the usual fully annealed material to enhance the strength of the gutter lining. Stop-end expansion joints are provided at high points between outlets, and the gutter is joined or weathered to the roof sheets and flashings by 'sliding' welts or overhanging aprons. The permissible length varies with the thickness of copper used, the width of sole, and the shape of the gutter section. As an example, the overall length of a box gutter lining made of 0.6 mm thick (24 SWG) quarter-hard temper copper strip, with a 300 mm (1ft 0in) sole and an outlet placed in the centre of the run, would be 7 metres (23ft 0in). If the thickness of the copper was increased to 0.7 mm (22 SWG) the length would rise to 14 metres (46 feet), whilst a reduction in width to 200 mm (8ins) would add a further 3 metres (9ft 10ins). Conversely, an increase in the width of sole to 400 mm (1ft 4ins) would produce

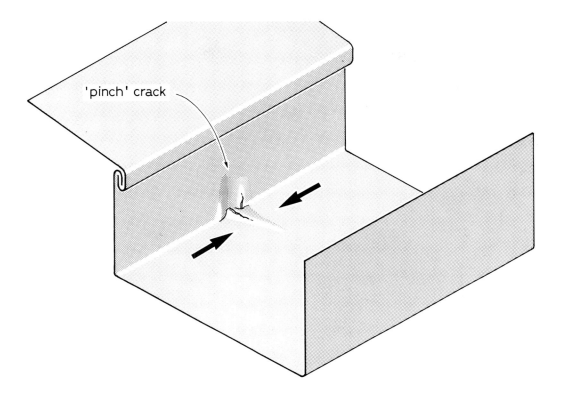

'pinch' crack

Figure 3.13   Typical failure due to excessive thermal movement in long gutters ('pinch' crack)

a permissible overall length for a gutter lining fabricated from 0.7 mm (22 SWG) thick copper of 12 metres (39ft 6ins approximately).

With stop-end expansion joints between every length of lining each section of gutter must be drained separately. It is not always possible or desirable to install additional rainwater pipes in or on old buildings to meet this requirement, or to move the existing ones to new positions suited to long-length gutter design. It is, in fact, this sort of situation which creates the problem in the first place.

The ability of double-lock cross welts to absorb longitudinal expansion movement in strings of bay sheets is well established. The main problem with using them for this purpose in gutters is that the welts are not watertight when subjected to a head-pressure of even a few millimetres, a service condition likely to occur every time there is a heavy downpour. Furthermore, where the fall in a gutter is minimal, 'ponding' is likely to keep rainwater in contact with welts for extended periods, during which time water is siphoned or drawn by capillary atraction through the welt. Sealing the welts with boiled linseed oil or non-hardening mastic compounds will prevent this, but only until the oil or mastic compound dries out, when seepage will commence. While at the initial stages the amount of

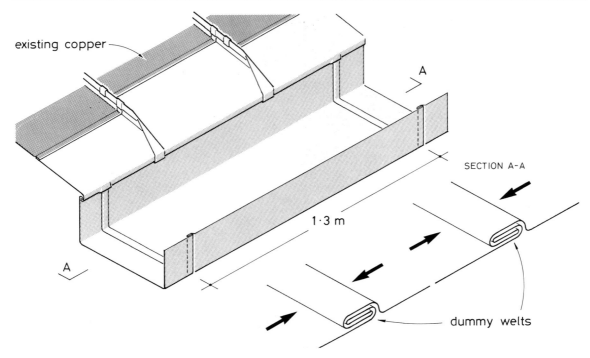

Figure 3.14   The formation of 'dummy' welts in a long gutter

water penetrating a double-lock cross welt may not be enough to cause a noticeable stain or damp patch inside the building, by the time the trouble has been observed and investigated boarding and roof timbers are often extensively damaged by wet rot.

The answer to the seepage problem is to reproduce the folds of a double-lock cross welt in a long strip of copper without actually cutting it into separate pieces, that is, to form 'dummy' welts across the gutter (Figure 3.14). A spacing of 1.3 metres (4ft 3ins) maximum for the 'dummy' welts will provide adequate accommodation for the thermal movement of the gutter lining between drips. Continuous runs of gutter up to 6 metres (18ft 0ins) between drips have been successfully dealt with using this method.

It will be necessary to cut out the 'worked' ends of the existing bay sheets which joined the defective gutter lining. The method of preparing and welting in new short bays to the existing covering is similar to that described for fitting an angle-fillet expansion joint, p.62.

Restriction of thermal movement — the use of incorrect felt underlay
All copper roofs must be laid on a suitable felt underlay to lessen abrasion with the decking, help deaden sound of wind and rain, prevent galvanic corrosion between the copper and ferrous fixings in the boards, and to smooth out any slight unevenness in the top surface of the decking. An inodorous sheathing felt conforming

to BS 747, or similar, is specified in CP 143 for copper roofing. It is vital that the felt should not adhere to the copper or understructure under temperature changes.

Although felt underlay performs the above functions admirably, it is not waterproof and this, combined with the need to lay the felt with butt joints to avoid ridges showing in the copper, renders it almost useless as a temporary weathering. This encouraged uninformed use of bitumen felt as an underlay during the 1950s, and as a result a number of copper roofs failed and were replaced.

The copper sheeting, having become thoroughly bonded to the substructure, is not only denied the freedom to move on its own account during temperature fluctuation, but is subjected to added stresses imposed by thermal movement in the roof structure. The copper sheeting is no match for the much stronger supporting structure and so is forced to bend and buckle at points where structural movement tends to be concentrated, i.e. seasonal and shrinkage gaps in the decking. The trouble manifests itself in a number of ways. Creases and buckles form in the metal and eventually develop into cracks and leak. Some double-lock cross welts crack internally and also finish up leaking. The 'bubbling' or, on steep pitches, sagging of the bitumen felt under the influence of heat absorbed by the copper on bright sunny days causes the centres of bays to bulge away from the decking and become unsightly.

Apart from the obvious visual signs which attend such defects, an early indication of the use of incorrect felt is the 'cracking' sound produced by walking on the copper. Once the roof is laid, there is nothing which can be done to delay or mitigate the consequences to the copper of the use of bitumen felt, but it must be pointed out that, although bulging in the bays takes place in nearly every case, cracking or perforation of the copper does not always follow. The shape, size and construction of the roof, plus its exposure and orientation to the sun, make each roof unique and its fate impossible to predict with any degree of certainty. There are still copper roofs in existence, laid on bitumen felt, which have survived for over thirty years, but whether it is the copper or the felt which is keeping the water out is difficult to say.

The only remedial treatment for seriously damaged roofs is to strip them and renew the covering. The 'bonding' effect of the bitumen makes even minor repairs extremely difficult. If a roof is not giving trouble, perhaps the best advice is to leave it alone but to review the situation from time to time.

Accidental damage
Falling masonry, scaffold poles and other objects are responsible for inflicting damage to many roofs at some time during their life. Damage is mostly of a minor and localized nature, and in the case of a fully supported traditional copper roof is usually no more than a shallow indentation in the metal and supporting boards, with perhaps a small rupture in the covering at the base of the depression.

(1) Small repairs A small hole is easily repaired with a copper patch cut to the required size and shape and soft soldered over the defect as shown in Figure 3.15. Somewhat larger repairs, involving removal of the damaged piece of copper sheeting and levelling of the indentation in the decking with a suitable wood

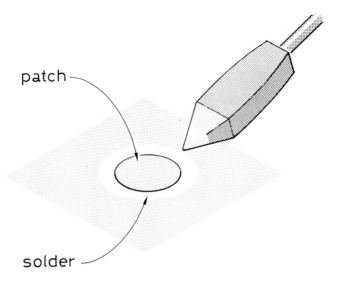

patch

solder

Figure 3.15   The soldering of a small repair

filler, are shown in Figure 3.16. The new piece of copper is silver brazed to the existing bay sheet using a 'dog-tooth' joint to hold the edges together and prevent undue distortion during brazing. Silver brazing alloys melt at temperatures ranging from 630 to 740°C (1164 to 1364°F), so the method can only be used where the copper can be raised from the boarding (which involves undoing a nearby seam), allowing a fire-resistant insulation sheet or pad to be placed between the two. If this is not practicable, a new square of sheet copper can be welted into the existing damaged bay, and the welt sealed by flowing soft solder under the final fold and into the mitred corners using a large copper bit (Figure 3.17). The edges of the new and existing copper must be pre-tinned and the welt dressed tight to create a capillary soldered joint for maximum strength.

(2) Large repairs Where the damage extends over one or more bay sheets, possibly involving broken boards, the only satisfactory course to follow is to remove the damaged sheets carefully, allowing the decking to be repaired before reinstating the felt underlay and closing the covering with new copper.

In view of the high cost of such repairs and the visual disturbance caused to an otherwise uniform green surface, it makes economic sense to 'board over' those sections of a copper roof most at risk from dropped or thrown objects during the time a scaffold is in place for repair or maintenance work on other parts of the building above or overlooking the copper roof. Replacement sheets may take as long as twenty years to weather in.

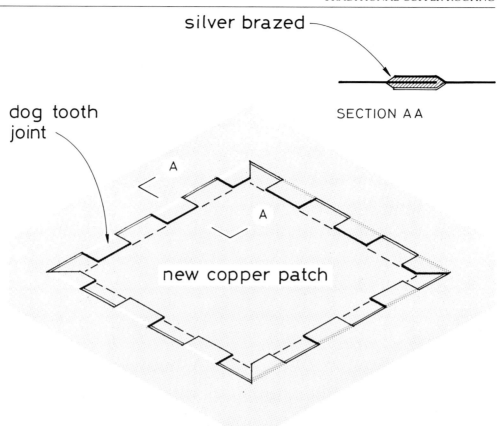

Figure 3.16   The silver brazing of a new copper patch using a 'dog-tooth' joint

Corrosion attack

Acidified rainwater run-off from other surfaces
The long life attributed to a copper roof is accounted for by the fact that copper develops by natural processes a surface film or patina which forms a protection against corrosion by the action of the atmosphere. In circumstances where rainwater becomes contaminated by corrosive agents in the atmosphere, or acids derived from organic growths on other roofing materials, the patina may break down or, indeed, be prevented from forming. If this should happen, instead of the long life expected, a client could be faced with premature failure of the covering in as little as five years.

(1) Mosses, lichens and algae on slates, tiles and duckboards Corrosion attacks occur most commonly in copper valley gutters draining rainwater from tiles, slates and asbestos cement sheeting covered with mosses or lichens. Dormer windows and

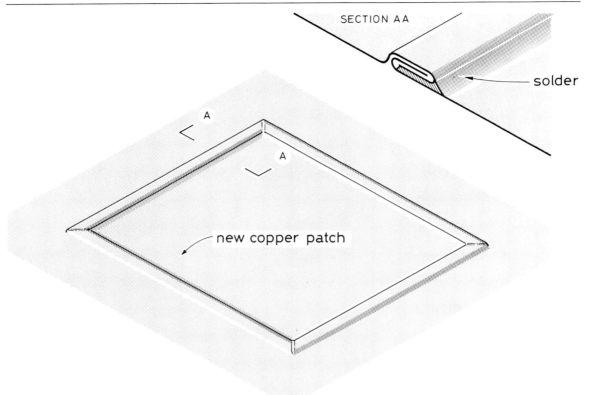

SECTION A A

solder

new copper patch

Figure 3.17   Welted and soldered copper patch

flat roof coverings which receive drips from tiled or slated slopes can be similarly affected where growths are sufficient to reduce the pH level of the rainwater sufficiently for it to dissolve the copper oxide which forms in the first stages of weathering, and even remove a well established patina. Renewal and dissolution of the surface oxide progressively thins the metal at the drip points, and unless remedial action is taken perforation will inevitably occur (Figure 3.18). The problem is easily identified by the characteristic 'rusty' looking stain marks on the copper at the drip points. Rubbing with firm finger pressure quickly removes the soft brown coloration, and confirmation of acid attack is obtained by the bright copper surface revealed.

There are several ways of dealing with this type of problem, but in all cases of corrosion attack the extent of damage to the copper must first be assessed before appropriate repairs can be carried out. Where the copper has perforated, a small area of sheeting immediately surrounding the hole will be very thin. The copper patch to be soft soldered over the hole should be large enough and shaped to cover this area. The copper surfaces to be joined must be pre-tinned with a copper bit to avoid the risk of fire to exposed supporting board and felt underlay. The patch and holed sheet are 'sweated' together with continuous pressure applied to the patch

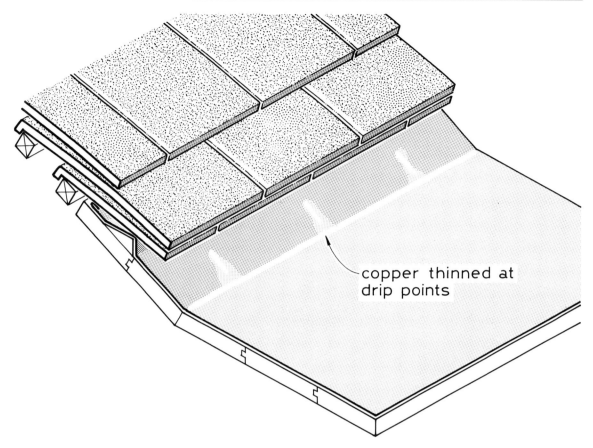

copper thinned at
drip points

**Figure 3.18   The effect on copper of acidified rainwater run-off due to lichens and
mosses**

during the soldering operation with a flat piece of fire-resistant material to obtain
a high strength capillary joint.

   If the attack is caught in early stages the copper will not have thinned to any
significant extent, and provided that the cause of the attack is eradicated (see p.73)
no other action need be taken and the corroded area will simply weather in the
normal way. Where thinning has continued for rather longer but has not yet
reached the point of perforation, the thinned spots may need patching to return
the covering to good order. A reliable 'rule of thumb' method of testing this is to
apply finger pressure on the thinned area. If the copper offers reasonable resistance
to indentation it will be satisfactory, but if it indents easily with minimal pressure
it would be prudent to solder a patch over it.

   In severe cases of corrosion attack the entire length of roof slope upstand to a
valley or parapet gutter may be so badly holed and thinned that a strip containing
the defective metal must be removed. A valley gutter is likely to be affected on

71

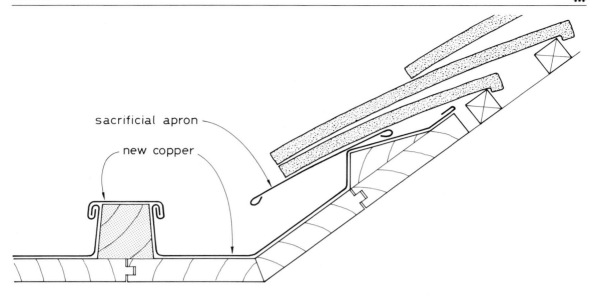

sacrificial apron

new copper

Figure 3.19   Installation of a 'sacrificial' copper apron to react with the acidified rainwater and protect the copper gutter

both sides and there would be no real alternative to renewing the gutter lining. On the other hand, if the valley gutter is very wide, perhaps divided into bays with batten roll joints in the manner of a small flat roof, or the gutter is behind a parapet with only one side under the tiles or slates, a repair similar to that shown in Figure 3.19 might be more appropriate and economical. In this instance a section of copper, including the damaged roof slope upstand and part of the gutter sole, is cut out to allow a new wood-roll to be fastened in the length, care being taken to allow sufficient width on the roof slope side to prevent blockage and flooding. The existing gutter sole is folded and dressed to fit against one side of the roll, and the new strip fits against the other side with a new capping to complete the weathering.

A 'sacrificial' apron is tucked below the under-eaves tile (Figure 3.19). The purpose of this apron is to react with acids in the rainwater and in the process reduce the potential of the rainwater to do harm to the copper gutter. In time the apron is destroyed, but it is a comparatively simple and inexpensive operation to remove it and insert a replacement.

An alternative method of protection is to paint the affected areas with two coats of bituminous paint. The main disadvantage of this treatment is the visual contrast to the pale green patina, and the fact that the treatment must be repeated at two- to three-year intervals in order to afford continuous protection.

Algae on timber duckboards and walkways can also prove harmful to a copper roof. It is not clear to what extent the timber itself may contribute to the problem, but timbers such as oak, columbian pine and red cedar are known to have distinctly acid pH values. Except when used in large areas, such as weather-boarding, or in the case of oak and cedar as roof shingles, the timber on its own does not appear to pose a threat to a copper covering. With the growth of algae on duckboards and walkways, rainwater can become sufficiently acid to attack copper.

Periodical treatment of the timber with a suitable algicide will take care of the growths, and bituminous paint applied to the copper beneath the sections of the duckboards and walkways, generally the more shaded parts, will give added protection. It is good practice, incidentally, to place bearing/protective pads on the copper sheeting to take the supporting posts or ribs of frequently used duckboards and walkways. Pads formed of hot applied bitumen reinforced with sisal or hessian will prevent any abrasion beneath the posts and at the same time afford protection against corrosion by algae. Provided the pads are small the bitumen is unlikely to harm the copper.

(2) Acid water draining from glass Localized corrosion, thinning and perforation of a copper roof can occur where rainwater drips from large areas of glass in roof lights and lantern lights. The rainwater becomes acidified by air pollution deposits which settle on the glass, and corrosion attack can be particularly aggressive in urban and industrial areas where solid fuel or heavy fuel oils are burned.

Visual evidence of attack, methods of repair and protection against further attack are similar to cases involving mosses and lichens. The sacrificial apron method of protection will be particularly suitable for those copper roofs where the glass occupies a prominent position and the obtrusiveness of bituminous paint would be aesthetically unacceptable.

(3) Treatment to kill mosses, lichens and algae It is something of a paradox that a metal harmed by mosses, lichens and algae is in turn lethal to such growths, but such is the case with copper. The fungicidal properties of copper are well known and many proprietary biocide solutions are based on or contain copper salts.

Evidence of the power of copper to kill mosses and lichens can be seen wherever telephone wires cross a roof, or where copper flashings are used around a chimney or skylight. The tiles or slates downstream of the copper flashings, or directly under the telephone wires, will be completely clear of mossy growths or lichens and will have a 'scrubbed clean' appearance.

It follows that a permanent answer to the moss and lichen problem is to install a 'copper deterrent' in the form of a strip of copper 75 mm (3ins) wide, tucked under the tiles or slates at about 5–6 course intervals. Only 12 mm (½in) of the strip need be showing, as rainwater will penetrate the joints between the exposed sections of the tiles, pick up copper from contact with the strip under the tiles, and re-emerge to wash over the surface below.

A solution of 1 part (100 g, 4 oz) copper sulphate crystals in 10 parts (1 litre, 1¾ pints) water has proved a suitable solution for killing mosses, lichens and algae. The crystals are slowly added to the water and stirred until dissolved. The solution is then placed in a knapsack spray and the roof material and growths are thoroughly wetted. The surplus solution should be allowed to drain from the roof. Rainwater gutters and pipes made of cast iron, steel or aluminium should then be flushed with clean water. This treatment should be repeated every three years. (See Volume 1 for treatment of organic growth on masonry.)

Concentrated flue gases

The problems associated with corrosive flue gases have largely disappeared with the replacement of solid fuel and oil-fired boilers by natural gas-burning boilers. Where oil firing has been retained, conversion from burners using high viscosity sulphur containing oils to low viscosity non-sulphurous oils has left few copper roofs still at risk.

Where products of combustion containing sulphurous gases and particles are discharged close to a copper covering and it is not possible to convert to a cleaner fuel, the replacement bays and affected zone around the chimney stack or flue, typified by a pitted and 'yellowed' appearance of the patina, should be painted with a minimum two coats of bituminous paint, and the treatment repeated as often as required to maintain a continuous coat. Alternatively, the stack or flue could be raised in the hope of dispersing the gases and particles clear of the copper roof.

Unfortunately, neither treatment is really satisfactory. Painting is unsightly, and very expensive if scaffolding is involved, and the flue can seldom be raised as high as required to disperse the gases efficiently, as this would spoil the appearance of the building.

Deterioration of understructure

Wood-boring insects and rot

Where copper is laid over timber it has been found that wood-boring insects such as deathwatch and furniture beetle are not attracted to the timber. This is attributed to the fact that metallic salts formed by condensation on the underside of the copper permeate the timber decking and create an inhospitable environment. Away from the decking, however, structural members of a timber roof are as much at risk from wood-boring insects as they would be in a roof covered with any other material. Similarly, the presence of copper in rainwater leaking through a copper roof, or in condensation formed on the underside of the covering, tends to discourage dry rot in the decking and timbers immediately beneath, but will not prevent them succumbing to wet rot.

Deterioration of the understructure of a copper roof may at times be repairable from inside the building or roof space without disturbing the covering. Replacement of certain supporting members such as purlins, struts, collars, and hangers can usually be accomplished without removing the copper, but rarely can a decking be repaired without first stripping the copper from the affected area.

Condensation

Moisture and heat introduced to a building as the result of human occupation give rise to water vapour, which rises to the highest parts of the building it can reach – usually the underside of the roof. Here it will collect, and unless provision is made to give adequate ventilation of the space or voids beneath the decking it will condense under conditions of heat loss, resulting in mysterious leaks and deterioration of the supporting structure.

In order to minimize the effects of condensation under a copper roof, inlet vents

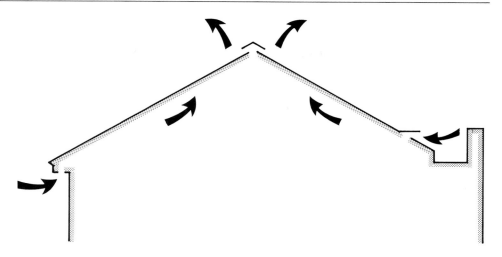

Figure 3.20   Providing natural ventilation to the underside of a copper roof

for natural ventilation of the roof space should be located at eaves level on both sides of the roof, with outlet vents at the ridge (Figure 3.20). When dealing with a flat roof, the vents should be positioned to provide cross ventilation of the voids beneath the decking. New copper roofs can be designed to incorporate the necessary inlet and outlet vents within the normal copper roofing details. With existing coverings this may not be possible, but sufficient 'eyebrow' or cowl-type vents constructed of copper can be installed at strategic parts of the roof, with air bricks in gable end walls, to ensure there are no troublesome stagnant zones. In particularly difficult situations it may be necessary to provide mechanical ventilation for the roof space, in which event specialist advice should be sought.

3.8   ARTIFICIAL PATINATION

Is it better to wait?

The development of natural green patina on a copper roof can take from ten to twenty years in the United Kingdom, depending upon the location and the composition of the atmosphere. This has led to attempts to produce a patina artificially in order to have a green surface finish from the outset. Although the methods tried can be successful with relatively small objects, none can be applied to large areas of sheet copper on a roof with any assurance of success.

Until more reliable artificial patination processes are developed which are not susceptible to the vagaries of the British weather during application and reaction periods and will not flake during service, or will permit copper roofing sheets to be commercially pre-patinated and then to be worked without significant loss of patina, informed opinion is strongly in favour of allowing a copper roof to develop its patina naturally.

75

Artificial patination formulae

Small copper patches and strips can be treated to assist harmonization with an established covering. Details of two chemical processes published by the Copper Development Association in their booklet No 57 *Copper Roofing* (out of print) are reproduced here by kind permission of the Association.

Formula No. 1 (British Patent No. 697294)

The solutions used contain a double salt of copper and ammonia but no free acid. The double salt, copper ammonium chloride ($CuCl_2 2NH_4Cl.2H_2O$), produces a pure green patina, but the addition of ammonium sulphate ($(NH_4)_2SO_4$) to the solution gives the true blue-green shade of the naturally formed patina. Thus the shades of patina can be varied within certain limits by altering the proportions of the salts, the ammonium sulphate controlling the degree of blue coloration. The recommended solution should contain 20 per cent by weight of copper ammonium chloride and 10 per cent by weight of ammonium sulphate, with the remainder water; the composition being as follows:

Copper ammonium chloride ($CuCl_2 2NH_4Cl.2H_2O$)	6oz	(170 grams)
Ammonium sulphate ($(NH_4)_2SO_4$)	3½ oz	(100 grams)
Water	1¾ pint	(1 litre)

Enhanced results may be obtained by adding to the solution small quantities, of the order of 0.02 per cent, of an organic silicon compound (a silicone) or trimethylhexanol, which have the property of assisting the retraction of the liquid into small globules on application to the metal surface, thereby achieving a more even effect. These materials are obtainable from most laboratory suppliers.

Preparation of the surface
The surface of the roof should be wiped clear of dust and dirt with clean brushes or cloths and greasy patches should be removed by swabbing with a solvent. The roof should be dry before application of the solution.

Application
It is suggested that this solution should be applied during a spell of fine weather. Application is best effected with a paint spray gun to produce a fine dispersion of the liquid at a concentration of 1 pint to 5 square yards (approx ½ litre to 4m²). At this concentration the surface should be covered with a uniform layer of droplets. This concentration should be varied at the discretion of the operator, so as to compensate for the action of the wind, but care must be taken to avoid undue coalescence of the droplets, which may occur if too much liquid is applied.

After application, it has been found essential that the surface should be free from any appreciable wetting until a certain amount of weathering has taken place. Premature wetting causes the patina to flake and peel. Atmospheric humidity undoubtedly plays an important part in patina formation. It is advised that where this factor is controllable, the best results are obtained with a relative humidity of between 85 and 100 per cent; although humidities as low as 65 per cent may be used, provided the period of weathering is lengthened accordingly. Alternatively, if the prepared surfaces are faintly damped by a fine mist or spray of water, good results may be obtained. The technique is to spray the prepared sheet with water so that the damped surface will dry out within a very short period of time, say five minutes, and this procedure is continued at intervals of not less than two hours between dampings until the requisite patina has been formed.

If the artificial patina is applied under normal outdoor conditions, it may well be impossible to follow directions concerning the control of humidity. However, we recommend that the prevailing and forecast weather conditions be studied before undertaking this process, avoiding where possible periods of heavy rainfall.

Formula No.2

Preparation of the surface

After all the work is completed, and before application of the colouring process, the copper surface to be coloured must be clean. Any dirt, oil or grease on the surface will prevent the necessary chemical attack incidental to the colouring treatment. The surface of all copper sheets is coated with a thin residual film of oil from the rolling operations. The surfaces are also soiled and spotted from handling and forming operations during installation.

The oil and accumulated dirt are generally sufficiently removed by the washing from two or three rainstorms, and experience has indicated the surface to be then in the best condition for colouring. However, if colour must be applied immediately after erection, the surfaces must be cleaned chemically by prepared cleaners especially suitable for removing oil, grease and dirt from metal, and these cleaners are available at hardware and paint stores. Cleaners of the trisodium phosphate type have been found suitable. In using cleaners, care must be taken not to leave the copper with too heavy a film of oxide.

Thorough washing of the copper after cleaning is essential, since any cleaning compound left on the surface will prevent proper chemical attack. Whether the surface is suitably cleaned from oil may be judged by the action of the droplets from the spray. The droplets on striking the surface should wet the surface and spread. If droplets do not wet the surface but tend to remain spherical and not spread out, the surface is not freed from oil and should be cleaned further.

It has been found that relatively thick oxide films will cause poor adherence of the patina. This is particularly true on roofs that have weathered for about six months to a year, depending on location. Colour will not develop satisfactorily on such roofs.

For such conditions it is necessary to remove the oxide film by washing with a 5 per cent to 10 per cent solution of sulphuric acid. The solution is applied cold. This is best done by swabbing with a cloth moistened with the acid solution, followed immediately by washing with clear water. The acid must be handled with care. Rubber gloves should be worn. Care must also be taken not to let any acid come into contact with adjacent wood or stonework. Treatment with sulphuric acid leaves the copper in excellent condition for colouring. It may be used to advantage on new as well as old roofs.

Application

Dissolve 1 lb (450 g) ammonium sulphate in 7½ pints (4½ litres) of water. Add ½ oz (14 g) copper sulphate and stir till it dissolves. Then add slowly ¼ fluid oz (7 cc) concentrated ammonia solution with constant stirring. The solution is now ready for use.

The method of application described below gives the best results. The solution is applied by spraying. An ordinary garden insecticide sprayer made of plastic will be suitable. The spray must be applied rapidly in such a manner that after spraying the solution rests on the copper surface in tiny isolated drops. It is better to apply too little than too much, since the excess will cause large drops to accumulate which will run down over the surface and cause streaks. After spraying the first time, the drops should be allowed to dry. On a warm, sunny day this requires about ten to fifteen minutes; on a cloudy day a little longer.

This spraying and drying period is then repeated (in the manner described) until five or six applications of the solution have been made. The colour does not appear immediately. Upon completion of spraying treatments, the surface of the metal should be covered with a glassy coating, in appearance somewhat like a dark coat of varnish.

The development of colour then depends upon suitable weather conditions. Rain falling on the roof within six to eight hours may spoil the work, washing off the salt before it has had an opportunity to act upon the copper. The ideal condition following an application of spray is a moderate to heavy dew. A light mist or fog is equally satisfactory, or any similar condition of relatively high humidity. It must be above 80 per cent relative humidity for chemical attack to occur. The salts deposited on the copper by the spraying are hygroscopic, that is, they absorb moisture from the air, which promotes the chemical attack on the copper. To obtain satisfactory depth of colouring the attack should proceed for about six hours. Provided this period of attack has occurred, the next rain will wash off the excess salt and bring out the blue-green colured patina. The colour that develops at first is somewhat bluer than the natural patina, but on weathering the colour approaches the natural shade. The time required depends upon climatic conditions.

REFERENCES AND FURTHER INFORMATION

References

1 Atkinson, R L, *Copper and Copper Mining*, Shire Album 201, Shire Publications, Aylesbury, England, 1987.
2 British Standards Institution, BSCP 143, *Sheet Roof and Wall Coverings, Part 12: 1970 Copper; Metric Units*.
3 British Standards Institution, BS 2870: 1960, *Rolled Copper Sheet, Strip and Foil*, Metric.
4 British Standards Institution, Bristish Standard Code of Practice 3, Chapter V, Part 2, 1972.
5 Copper Development Association, *Copper in Roofing – Design and Installation*, TN 32 CDA, Potters Bar, December 1985.

6 Copper Development Association, *Copper Through the Ages*, CDA Publication No 3, first issued 1934, revised 1956.

7 Glover, H, 'The Development of Light Gauge Sheet and Strip for Roofing and Cladding', paper given at *Sheet Roofwork and Weathering Symposium*, conference organized by the Instiutute of Plumbing, Oxford College of Further Education, 1 April 1987.

8 Hemming, D C, 'The Production of Artificial Patination On Copper', *Corrosion and Metal Artefacts — A Dialogue Between Conservators and Archaeologists and Corrosion Scientists*, National Bureau of Standards Special Publication 479, July 1977, pp 93–102.

9 Institute of Plumbing, *Sheet Roofing Data Book and Design Guide*, Technical Committee, Oxford District Council and the Institute of Plumbing, Oxford, 1978.

10 Sivinski, Valerie, 'Conservation Handbook for Copper and Copper Alloy Architectural Ornamentation,' MA in Conservation Studies, University of York, Institute of Advanced Architectural Studies, August 1986.

Further information

Copper Development Association
Orchard House
Mutton Lane
Potters Bar
Herts EN6 3AP
Tel: (0707) 50711

British Non-Ferrous Metals Federation
Crest House
7 Highfield Road
Edgbaston
Birmingham BL5 3ED
Tel: (021) 454 7766

4 THE REPAIR AND MAINTENANCE OF OUTDOOR BRONZE SCULPTURE*

4.1 UNDERSTANDING THE PROBLEM

Traditionally bronze used for statuary is an alloy of about 90 per cent copper and 10 per cent tin, although zinc, lead, aluminium and silver may also be present in very small quantities. It is the best known and most widely used alloy of copper. Bronze is a salmon-gold coloured metal which is seldom seen without the dark brown-red of the oxidized surface or the green patina which develops in urban, marine or industrial atmospheres. Bronzes were used for casting from prehistoric times, although the nineteenth-century was the peak period for urban bronze sculptures. Many nineteenth-century bronzes were treated with heat and/or chemicals to produce a deep brown patina which was a part of the original artistic intention.

The conservation of a bronze sculpture in an outdoor situation requires detailed understanding of both the object and the corrosion phenomena which are affecting it. A bronze sculpture is a highly complex fabrication which must be treated sensitively. Relevant to its conservation are an understanding of factors such as methods of manufacture, metallic grain structure, alloy composition, the extent of flaws and patches, the nature of the interior support, and the types of surface finish and patination. These factors will influence, in varying degrees, the way in which a bronze reacts with its environment.

The nature of the environment of a bronze must also be understood and there may be a wide range of variables which will affect the rate and degree of

* Thanks are due to Dr Barry Knight and Mrs Marjorie Hutchinson of the Ancient Monuments Laboratory for their assistance in the preparation of the text.

deterioration. The most significant of these will be moisture, particulate matter and sulphur dioxide, largely a modern, post-industrial phenomenon.

4.2 THE PRIME CORROSIVE AGENTS

The three prime corrosive agents, moisture, gaseous sulphur compounds and particulate matter, work with other gaseous pollutants and oxygen to establish un-desirable surface conditions on bronze. These atmospheric agents work on the basis of a combination of chemical and electrochemical reactions.

Table 4.1 indicates the commoner active agents and the colour and nature of the salts which are found when they attack bronze in the presence of moisture.

Table 4.1
The effect of corrosive agents on bronze

Active agent	Salt formed	Colour
Oxygen	Oxide	Red-brown to dark brown and black
Oxygen and chlorine	Chloride or oxychloride	Very pale green
Carbonic acid	Carbonate (malachite)	Green
	Carbonate (azurite)	Blue
Sulphuric acid	Sulphate	Deep blue
	Basic sulphate	Green
Nitric acid	Nitrate (soluble and deliquescent)	Blue-green to blue
Sulphur	Sulphide	Dark brown or black

(Reference 3, p. 231)

Unlike with copper, the green patina on bronze is usually not protective.

Particulate matter may be inert particles of dirt and dust or it may be particles which contain acidic, bituminous or other corrosive substances. The particles are deposited in dry form as dust or they arrive suspended in rainwater. Particulate matter can act as a catalyst to corrosion and it is desirable to remove it so that electrical corrosion cells are not set up. This is one of the reasons wax coatings on bronze to which particles adhere readily need to be removed and replaced frequently.

In an urban, polluted environment corrosion will proceed far more rapidly than in a more rural environment, where the potency, quantity and nature of the corrosive agents is less damaging.

In a highly polluted, urban environment, surface corrosion may appear rapidly in the form of deep, localized pitting and bright green and black corrosion products. Localized electrochemical corrosion cells usually develop next. Areas covered with dark corrosion products act as cathodes (protected) and those covered with bright green corrosion products act as anodes, under which the most serious loss of metal occurs. The black areas stand proud of the green. During

Bronze sculpture in an urban environment develops a patina containing several corrosion products many of which are unsightly and encourage further decay. However, sometimes this patina will contain remains of the original surface colouring or possibly even gilding. Following analysis, bronze sculpture should be cleaned and treated in the workshop of a metals conservator.

the formation of both the green and the black corrosion layers, the original patina and the surface of the metal are destroyed. If the corrosion process is allowed to continue, disfigurement will result.

Particles of dirt lodge on both types of corrosion layers. Those that accumulate on the green layer, which is highly soluble, mostly wash away. On the black layer the particles are usually retained and incorporated in the crust. When a particle lodges on the surface of bronze, conditions are created beneath it which are suitable for an electrolytic action cell, that is, corrosion.

The shape of a statue will strongly influence its weathering pattern. Water which collects in crevices will become increasingly corrosive as it evaporates. Unwashed areas are more likely to show serious corrosion than regularly washed convex areas. Unsightly streaks and stains may be the effect of rain on deposits of dust, soot and tar. The degree of corrosive attack of bronze relates directly to the length of time it remains wet/moist, which again relates to shape.

Two other forms of bronze deterioration, which are unlikely to be experienced often in outdoor bronze, are 'bronze disease' and dezincification.

'Bronze disease' most commonly affects bronze artefacts that were buried. It is seldom a problem with outdoor bronze sculpture or architectural elements.

Dezincification is a type of corrosion peculiar to bronze or brass with a composition of more than 15 per cent zinc. It occurs in the presence of acids and other strongly conducting solutions. The copper–zinc alloy is dissolved, the copper is redeposited electrochemically and either the zinc remains in solution or its compounds form a scale. The metal may be left pitted, porous or even weathered, depending on the severity of the dezincification.

Investigation of bronze statues

The conservation of a bronze artefact in an outdoor exposure requires analysis of the condition of the piece and the corrosion processes before appropriate conservation proposals can be made. It is important to recognize the difference of approach between architectural bronze elements and bronze sculpture. The inspection, assessment and remedial work on a bronze sculpture should be under-taken only by a metals conservator. It is understood that the Museums and Galleries Commission will have a list of metal sculpture conservators in the near future.

Several methods of non-destructive testing can be used to provide a detailed evaluation of the quality and condition of a bronze. Ultrasonic testing, radiography, X-rays, thermovision (thermal imaging) and acoustic emission can be used in conjunction to determine the homogeneity of castings, locate critically thin areas, verify the structural continuity of joints, show up voids in the casting, identify other defects such as previous repairs and locate plastic deformation and/or microcracks. An endoscope will enable a close visual inspection of internal surfaces to be made. While these methods can provide a lot of interesting and valuable information about the manufacture and condition of a bronze sculpture, they are not always readily available, affordable or practical. A detailed inspection of the piece by a person experienced in the behaviour and characteristics of bronze, possibly coupled with a simple analysis of corrosion products, may be all that is practicable and necessary. The historical value of the bronze will, of course, be a prime deciding factor.

The value of patina on bronze

The patina determines the appearance of bronze and has an aesthetic value by virtue of its colour and texture. The chemical composition of patina is determined by the composition of the metal and the influence of either reagents in the atmosphere or chemicals applied by man, or both.

In an environment without pollution, bronze develops a protective layer of cuprous oxide which is dark brown in colour. Today the air nearly always contains other agents which create other corrosion products of various colours. Basic copper carbonates form a stable layer which is blue and green. Copper chlorides, also blue and green, are a destructive and undesirable corrosion product. Sulphuric acid agents will produce a pale green patina, and hydrogen sulphide creates a black corrosion product. It must always be remembered that a bronze sculpture may have been artificially patinated by chemical or heat treatments in the works as part of its original artistic design. This may be excluded by or contained within corrosion products created in the atmosphere. The layer of patina may or may not protect the metal beneath.

The history of artificial patination on sculpture and objects of artistic value is over 2000 years old. From the nineteenth century onwards the artificial patination of bronze into colours such as rich chestnut brown, dark brown, black and many shades of green and blue was developed and subsequently practised on a large scale (reference 7, p17).

The issue of the removal or retention of patina is frequently debated in conservation circles. It involves identification of the original patina and subsequent changes to this, and a judgement on whether the corrosion products are disfiguring or destructive. Each situation needs to be judged on its own merits. The necessary investigations, assessment and decisions are the realm of the metals conservator.

4.3   CLEANING BRONZE

The cleaning of bronze may simply involve the removal and replacement of a protective coating. At the more severe end of the scale, it may involve the removal of corrosion products and patina, repatination and the application of a protective coating.

Making the decision to clean

Before the decision is made to clean a bronze sculpture or a cleaning treatment selected, the following questions need to be answered:

- What will the cleaning accomplish? Why is it necessary?
- What was the original designer's/manufacturer's intent?
- Can the corrosion be removed without too much harm and what will be left if it is removed?

- Why is the object being cleaned, how much damage is acceptable and what is the acceptable level of clean?
- What will it look like after it is clean?

- What limitations does the object impose on the cleaning?
- How big is it, can it be removed/dismantled, what materials are involved?

- How severe is the corrosion?
- What is causing the corrosion and can it be eliminated or reduced?
- How severe is the damage?
- What are the corrosion products and are they loosely or firmly attached?

- How will the cleaning affect the object's future performance?
- Will a protective coating need to be applied?
- What are the maintenance obligations and when should it be recleaned?

The removal of grease and dirt

This section considers the removal of dirt and grease from bronze as part of a maintenance programme, not the removal of corrosion products, which is considered in the following section.

Dirt on bronze is usually comprised of particles of dirt and other oily, greasy and tar-like substances. Particles of dirt can contain corrosive acids which will gradually pit the surface of metal beneath and cause staining. Even if the particles are chemically neutral, the oxygenated surface will corrode. The dirt will be easily removed and will have done little damage if it is frequently removed as part of a maintenance programme. In such situations cleaning can be achieved by washing with a neutral-pH soap in warm water, using a lint-free cloth or natural bristle brush (copper, brass, bronze or ferrous brushes should not be used). Washing should be followed by thorough rinsing, drying with another cloth and the immediate application of a thin protective coating. The soap may need to be diluted 1:1 with water or white spirit to remove greasy dirt. This simple method will not affect insoluble corrosion products or patina. Glass fibre brushes and wooden scrapers can be used to remove loosely adherent material such as bird droppings. Tenacious mineral deposits or excretions emanating from core material may be treated locally with gentle abrasion using fine bronze wool (not steel wool). Any protective coating would need to be repaired in such areas. In any cleaning operation small tools are recommended as these will give controlled and gentle scrubbing.

Where it is necessary to clean and de-grease only, all acid- and alkali-based solutions should be avoided as well as all abrasive agents. The same applies to the traditional use of ammonia solutions, which will remove patina and can form an orange coloured surface. The solvent trichloroethane is a successful de-greaser but it is highly toxic. The best approach is a non-caustic de-greasing gel, several of which are readily available. De-greasing should involve thorough washing and drying if it occurs after washing and before the application of a protective coating.

Between fifteen and thirty years ago bronze sculptures were often cleaned with highly acidic or alkaline solutions, some of which were not washed out thoroughly. The residual problem may need to be diagnosed and dealt with.

If the water used during the washing or de-greasing of bronze is hard, it is advisable to use distilled or de-ionized water in the final wash. It is always advisable to dry a bronze after a period of wetting to avoid water spotting. During this and any other method of cleaning, ladders, sheeting, scaffold poles and ropes must not be allowed to remain in contact with the bronze for periods of more than a day. Corrosion can easily be set up at points of contact, causing the loss of patina at least.

Should it be necessary to remove paint from bronze, a non-caustic, methylene chloride-based gel should be used. Acetone can usually remove marker pen successfully.

Pigeon droppings contain a high proportion of soluble salts. A layer of droppings produces a corrosive poultice beneath which severe corrosion will occur if it is not removed quickly. Water which runs off collections of droppings can produce very disfiguring streaking. Pigeon droppings should be removed from bronze surfaces as soon as possible with timber scrapers and bristle brushes, and the surface should be thoroughly washed.

The removal of surface layers

All the corrosion products on a bronze sculpture may need to be removed, for one of several reasons. The products themselves may be promoting further corrosion, they may be marring the aesthetic appearance of the bronze (spotting or streaking), or they may form a non-adherent layer which would provide an unsound basis for a protective coating.

The removal of surface layers, corrosive or protective, leaves the bronze beneath open to new attack and must, therefore, be considered in conjunction with the application or reapplication of a protective coating. The methods of removing surface layers will be considered first.

Air abrasive techniques

During the last decade small-scale air abrasive methods have been used for the cleaning of outdoor bronze sculpture. One of the best known methods is micro-blasting with smooth glass beads (75–125 microns) through an air abrasive pencil or a light duty suction gun at 80–100 psi (11–14 kPa). The surface of the bronze may be reworked or 'peened' by this method. Alternative air-driven abrasives which have a gentler action include crushed walnut shells and coconut shells. Finely crushed olivine and aluminium silicate are also used at times.

If these methods are used carefully by trained conservators, virtually no metal is removed. Air abrasive methods can be used selectively on the whole or on part of a bronze to remove all or only part of a corrosion layer. The skill of the operator and the choice of powder are critical. Larger-scale abrasive cleaning machines, higher pressures or more cutting abrasives should never be used.

Chemical methods

Acid or alkaline chemical cleaning solutions react with the corrosion products of bronze, breaking the bond with the metal so the products can be washed away. They can be used for an overall clean or in a poultice as a localized solvent for use on a specific area. Chemical cleaning systems are usually effective and can provide a clean, bare metal surface free of corrosion. If the corrosion was thick the surface will probably be pitted.

However, as a chemical cleaning agent does not differentiate between thicker and thinner deposits, etching can occur. This general lack of control can mean the loss of an original surface which is contained within the corrosion products. Furthermore, some chemical cleaning agents may leave new damaging residues and others are highly toxic. Because of this the sculpture must be thoroughly washed afterwards.

Chemical/abrasive methods

While chemical cleaning agents remove corrosion products by dissolving them, mechanical methods remove them by physical impact. Polishing liquids include both chemical cleaning agents and fine abrasive. They clean predominantly by abrasion, i.e. removal of metal surface. Although they are readily available the implications of their use should be carefully considered.

The removal of a thin green encrustation can be achieved with a mildly alkaline solution such as a nearly saturated solution of washing soda (sodium carbonate) in water (reference 3). This solution is not harmful to human skin and as the sodium carbonate is readily soluble any residue can be easily removed by thorough washing. A solution of 4–8 oz (125–250 grams) sodium metasilicate in one gallon (4.5 litres) of water is another mildly alkaline solution which achieves similar results (reference 5). Both solutions can be assisted by the use of finely powdered pumice and a felt pad. Pumice powder can also be incorporated in paraffin. Great care must always be taken when such an abrasive is used because the paste formed will obscure the surface under treatment. Frequent clearing of the paste will be necessary to check on the level of clean. A further solution is a slurry of 5 per cent oxalic acid and water in a clay paste and/or finely ground pumice powder applied with a clean, soft cloth (reference 3). Clean water rinsing must follow. For small objects the safest and most effective treatment in the removal of heavy encrustation is the use of one of the 'biological' corrosion removers from a supplier of conservator materials. Remember that, if the original patina is within the corrosion encrustation, it will be removed too.

An article by J.F.S. Jack (founder of the Laboratory of the Ministry of Public Buildings and Works), 'The Cleaning and Maintenance of Bronze' (1953), was a pioneering document in the field of bronze cleaning, but it is now known to suggest an approach which is severe and is no longer recommended. The article suggests that heavy or resistant encrustations be removed with a solution of 1 part strong ammonia to 1 part water, assisted by a wire brush or a piece of natural pumice, and followed by repeated washing. Jack also describes a paste of 1 part caustic soda, 2 parts lime putty and 2 parts sawdust which is left on for 3–4 hours and followed by repeated washings. Both solutions are highly corrosive and are unlikely to be recommended today.

Gentle tapping with the edge of a small hammer can loosen and fracture a thick, unwanted layer of corrosion products. The use of heat to remove encrustations should only ever be undertaken by a metals conservator, as inexperienced application can cause fracturing and colour changes.

Elecrolytic reduction

Electrolytic reduction is often limited to small artefacts. These are placed in a solution which carries an electric current and the corrosion products are converted back into the base metal. It is also possible to use the method locally on larger items which are not immersed. Its use is generally restricted to the museum field.

Selecting a method of cleaning

It is not possible to recommend one particular cleaning system, mechanical or chemical, which is the most appropriate in all situations. Any particular piece of bronze that needs cleaning will have its own requirements, limitations and possibilities. These must first be understood and must form the basis for the selection of a cleaning system or combination of cleaning systems.

Pedestals, plinths and paving should be protected from all cleaning processes with plastic sheet and tape on battens pinned into masonry joints with non-ferrous or galvanized nails.

4.4 SURFACE TREATMENTS FOR THE PROTECTION OF BRONZE

It has been known for some time that the application of a surface coating such as wax or lanoline provides a barrier which prevents the corrosive impurities of the atmosphere coming into direct contact with bronze. Once a bronze surface has been cleaned and stabilized this is still the best approach to the conservation of bronze surfaces.

Today, the surface coatings include lanolin, traditional natural waxes, synthetic waxes and acrylic lacquers. At times a corrosion inhibitor is added. While some coatings may last slightly longer than others, all should be considered as preventive maintenance treatments which require repair or renewal to remain effective.

Traditional treatments

Traditional surface treatments for bronze have been based on lanolin or beeswax.

A satisfactory form of lanolin solution consists of 40 per cent lanolin, 7 per cent paraffin and 53 per cent white spirit. This coating was used in England from 1950 to about 1970. It was recognized that this coating could provide only a temporary barrier, as it had some ability to absorb water containing corrosive agents. The coating was replaced 2–4 times per year, the old lanolin having been completely removed with white spirit, paraffin or synthetic turps.

It is important that only a very thin coating of lanolin or any other wax coating is applied. This reduces the likelihood of particles of dust or soot adhering, especially in unwashed areas, and forming points at which localized corrosion of the metal will occur.

Another traditional surface coating for bronze includes beeswax and turpentine. The London bronzes which are under the care of the Property Services Agency are cleaned and treated with a mixture of lanolin, paraffin, beeswax and turpentine.

The conservation of bronze sculpture in outdoor situations requires an understanding of the material itself and its reaction with the environment. Many of the London bronzes are protected by a coating of natural waxes which is removed and reapplied at least on an annual basis and in this way retain their original artistic intent. The cleaning, repair and treatment of bronze sculpture is usually the realm of the metal sculpture conservator.

Those which are readily accessible are treated every three months, this being the desirable period of maintenance. The London bronzes are consequently in excellent condition, for which they have an international reputation. The original patination and artistic form of these works can be appreciated.

More recent treatments

In more recent years microcrystalline synthetic waxes have been applied to cleaned external bronze. They have a higher melting point than animal waxes. It is not possible to state categorically whether synthetic waxes offer better or worse protection than natural waxes, as there is evidence both ways. Both types require frequent removal and replacement.

In the early 1970s the acrylic resin-based lacquer 'Incralac' was developed by the International Copper Research Association for use on copper-based metals. The lacquer also contains ultraviolet inhibitors, plasticizers, hardeners and benzotriazole (BTA), a corrosion inhibitor. The lacquer is usually sprayed on, sometimes in a series of coatings. It produces a high gloss surface unless a matting agent is added. Incralac must be applied to a thoroughly cleaned and stabilized surface. As the high gloss surface is often aesthetically unacceptable, the lacquer is usually followed by a coat of resin or synthetic wax. 'Incralac' is only able to tolerate the dimensional movements of the bronze beneath for a limited period of time, after which it will begin to craze. A deteriorated coating of this kind can look worse than a deteriorated wax coating and can promote corrosion in the areas of breakdown. 'Incralac' should be applied by spray, in a thin coat. If it is applied by brush it is easy for accelerated corrosion points to set up at discontinuities in thin areas caused by brush hairs. An acrylic lacquer will probably need renewing at intervals of 1—4 years, depending on factors such as location, exposure and the level of preparation of the metal beneath. As with natural and synthetic wax coatings, a treated surface must be inspected no less than annually. The old lacquer must be removed by dissolving in either toluene (highly toxic) or methyl ethyl ketone.

In Canada a stabilizing wax has been formulated as an alternative to lacquers. It is based on a microcrystalline wax, toluene and acetone, and it includes the corrosion inhibitor benzotriazole. The coating has been used as an alternative to Incralac and is reported to be performing well (reference 4).

Long term protection

At present there are no coatings which will provide medium or long term protection to exterior bronze. The survival of any protective coating depends on the maintenance it receives. This should involve the washing off of dirt, dust and other corrosive deposits and a thorough inspection of the coating to determine that it is undamage and still firmly bound to the bronze. The visible signs of decay in a coating will be a change in colour to a slightly greenish or blackish tint or the appearance of small green spots. Even with intermediate repairs, all barrier coatings need complete removal and replacement at periods of up to 3 years.

Few materials benefit more from regular maintenance than bronze in outdoor conditions. Maintenance of bronze ensures its protection and restricts corrosion, it prevents distortion and falsification of the artistic message and avoids expensive, major conservation treatments.

4.5 THE REPAIR OF BRONZE

Repair of bronze is positively the field of the metals conservator. A dented bronze object may be hammered back in place if the backside is accessible and the material is not too thick. It must again be emphasized that this is skilled craftwork and that hammering can produce fractures. Severely damaged sections can be cut out, recast, and detached by riveting or brazing (a form of soldering with a bronze or brass filler metal), and fractures may be metal stitched. Missing pieces can be reproduced and fixed in a similar manner. Some scratches can be buffed to match the original finish and texture, as can brazed joints. New castings can be produced at one of the several bronze foundries which operate in the UK (contact the Association of Bronze and Brass Founders at the British Foundry Association in Birmingham). New castings must be metallurgically identical to the original, otherwise they will weather differently or promote corrosion.

4.6 REMOVAL OF GREEN STAINS ON MASONRY

Areas of masonry which receive the run-off from neglected bronze surfaces will most probably be stained green by copper salts corrosion products. Such stains can be reduced in intensity or removed by the application of a paste of 1 part ammonium chloride/aluminium chloride to 4 parts powdered talc. The stained area and surrounds should be thoroughly pre-wetted, the paste applied and left until dry. It should then be brushed off and repeated if necessary. After treatment the area should be washed thoroughly with water, as residues could cause soluble salt crystallization problems. If the stain has penetrated into the stone it may reappear and require further treatment. The metal must be carefully protected from contact with the paste.

It is always better to prevent bronze staining than to have to remove stains. The maintenance of a wax coating will prevent the problem occurring. The installation of gutters and drips should also be considered. The use of a water repellent on the stone plinth is generally not recommended in this situation.

4.7 REPATINATION

It is possible but not always easy to repatinate bright areas of bronze which have necessarily been revealed during cleaning. Various combinations of chemicals are used, depending on the required colour. A thorough reference on this field is *The Colouring, Bronzing and Patination of Metals* (Richard Hughes and Michael Row, Crafts Council, London, 1982).

Repatination is most definitely a specialist skill which requires time-consuming tests to make sure the right colour is produced. Repatination also begs the question of whether the metallurgical history of the patina is being interfered with.

REFERENCES AND USEFUL ORGANIZATIONS

References

1 Chase, W T and Veloz, Nicholas F, 'Some Considerations In Surface Treatment of Outdoor Metal Sculptures', The American Institute for Conservation of Historic and Artistic Works (AIC), preprints of papers presented at the thirteenth annual meeting, Washington DC, 22–26 May 1985.

2 Gayle, Margot, Look, David W and Waite, John G, *Metals In America's Historic Buildings – Uses and Preservation Treatments*. US Department of the Interior, US Government Printing Office, Washington DC.

3 Jack, J F S, 'The Cleaning and Preservation of Bronze Statues', *The Museums Journal*, Vol 50, No 10, January 1951, pp 231–5.

4 La Fontaine, Raymond H, 'The Use of a Stabilizing Wax to Protect Brass and Bronze Artifacts', *Journal of the International Institute for Conservation of Historic and Artistic Work – Canadian Group*, Vol 4, No 2, short contribution.

5 Sivinski, Valerie, 'Conservation Handbook for Copper and Copper Alloy Architectural Ornamentation'. MA in Conservation Studies, University of York, Institute of Advanced Architectural Studies, August 1986.

6 Smith, Rika and Beale, Arthur, 'An Evaluation of the Effectiveness of Various Plastic and Wax Coatings In Protecting Outdoor Bronze Sculpture Exposed to Acid Deposition: A Progress Report'. ICCROM Conference, *The Conservation of Metal Statuary and Architectural Decoration In Open-Air Exposure*, Paris, 6–8 October 1987.

7 Stambolov, T, 'Introduction to the Conservation of Ferrous and Non-ferrous Metals', *The Conservation and Restoration of Metals*, Scottish Society for Conservation and Restoration, proceedings of the symposium, Edinburgh, 1979, pp 10–19.

8 Walker, R, 'The Role of Benzotriazole in the Preservation of Antiquities', *The Conservation and Restoration of Metals*, Scottish Society for Conservation and Restoration, proceedings of the symposium, Edinburgh, 1979, pp 40–49.

9 Weil, Phoebe Dent, *Maintenance Manual for Outdoor Bronze Sculpture*, Washington University Technology Associates (WUTA), St Louis, Missouri, January 1983.

    This author has written prolifically on the corrosion, analysis, cleaning, repair and maintenance of outdoor bronze sculpture.

See also the Technical Bibliography, Volume 5

Useful organizations

Association of Bronze and Brass Founders
136 Hagley Road
Birmingham B16 9PN
Tel: (021) 454 414

British Foundry Association
Ridge House
Smallbrook Queensway
Birmingham B5 4JP
Tel: (021) 643 3377

British Non-Ferrous Metals Federation
Crest House
7 Highfield Road
Edgbaston
Birmingham BL5 3ED
Tel: (021) 454 7766

Copper Development Association
Orchard House
Mutton Lane
Potters Bar
Herts EN6 3AP
Tel: (0707) 27711

Institute of Corrosion Science and Technology
Exeter House
48 Holloway Head
Birmingham B1 1NQ
Tel: (021) 622 1912

National Corrosion Service
National Physical Laboratory
Teddington
Middlesex TW11 0LW
Tel: (01) 977 3222

5 THE REPAIR AND MAINTENANCE OF LEAD SHEET ROOFING*

5.1 INTRODUCTION TO LEAD IN ROOFS

Lead is the softest of the common metals and in a refined form is very malleable. It is capable of being shaped with ease at ambient temperatures without the need for periodic softening or annealing, since it does not appreciably work harden. Lead sheet can, therefore, be readily manipulated with hand tools without the risk of fracture. By the technique of bossing, lead sheet can be worked into the most complicated of shapes. Lead flashings can be readily dressed *in situ* to quite complex profiles such as moulded stone or deeply contoured roof tiles.

The main use of lead in historic buildings was in comparatively thick sheets as a roof covering, accompanied by lead gutters, flashings, downpipes and rainwater heads. Properly specified and detailed, the useful life of lead sheet is upwards of 150 years. It can be laid to any pitch, readily repaired *in situ* and recast. The traditional method of manufacture of lead sheet was by casting on an open sand casting table. This craft is still available. Milled lead sheet was widely adopted in the early part of the nineteenth century and may at times be considered inappropriate for repair work on many buildings. Lead is also used to make lead cames, the H-section strips which are soldered to form a substantial part of the support structure for the various glass shapes of 'leaded light' windows, and is cast in decorative sculptural forms (see Volume 5, Chapter 2 'The Repair and Maintenance of Historical Window Glass').

Lead alloyed with tin to form a protective coating on sheet iron or steel is known as 'terne coating'.

*RTAS is grateful to Mr Richard Murdoch for his constructive comments during the preparation of this note.

93

The malleability of lead makes it easy to boss it around irregular shapes to provide protection from the weather. It is a good investment to provide lead sheeting weathering to horizontal surfaces or protruding features which would otherwise decay rapidly.

Essential reference material

Current good practice in detailing is nowhere better and more graphically explained than in *Lead Sheet in Building* and a regular issue of technical notes produced by the Lead Development Association (see References). These are recommended as essential reading.

Occurrence and extraction of lead ore

The main lead ore, galena (lead sulphide, PbS) is usually found in veins, accompanied by other minerals such as quartz, calcite, fluorite and barite, and at times contains silver. It is found in all the highland regions of Britain and has been mined in most of them. Lead ore is particularly common to limestone, partly because that rock has suitable cavities in it, but almost any hard rock may contain it. Most of the major veins of lead ore are found in fissures naturally enlarged from faults or joints in the rock. They are therefore usually near vertical and can be traced for several miles to depths of 300 metres (1000 feet) or more. Sometimes the ore is found in layers between beds of rock, especially in limestone.

Because of its widespread occurrence and the ease of its smelting, lead was one of the first metals to be explored in Britain. It has been found in small pieces in pre-Roman burial mounds. Lead was widely used during the Roman period for pipes and cisterns and as sheet lead for roofing and baths (Bath and Buxton). Some was converted into pigment. With the departure of the Romans the need for lead and mining skills went into decline. In medieval times only relatively small quantities were used for roofing churches and some other buildings. A revival came in the late sixteenth century with the increase in building and rebuilding of great houses. The great age of British lead mining came in the eighteenth and nineteenth centuries, when Britain was the world's main producer and developed new techniques of mining and smelting. However, the importation of lead began in the 1830s as manufacturers found this cheaper than the local sources. In the early 1880s lead was even being imported from Australia, and the British lead mining industry declined rapidly. Only a few mines managed to continue into the twentieth century, the last of these closing in 1960. Today lead is mined in Britain only as a by-product of other materials.

5.2   AVAILABLE FORMS OF LEAD SHEET

Sand cast lead sheet

Cast lead sheet is still made as a craft operation by the traditional methods of running molten lead over a bed of prepared sand. A comparatively small amount is produced by specialist leadworking firms largely for their own use, in particular for replacing old cast lead roofs and for ornamental leadwork. The minimum thickness that can be sand cast is equivalent to a weight of about 7 lb per square foot (code 7), although in many cases much thicker lead sheet was used in the past. The long life of cast lead sheet was achieved partly because of the thickness used and partly because advantageous impurities such as silver, which gave mechanical strength, were often present. After the Industrial Revolution and the invention of

95

Lead sheet roofing has been used on both simple and complex roof forms. Provided the material was fixed properly in the first place it has a useful life in excess of 150 years. The high corrosion resistance of lead contributes to this performance. Traditional sand cast lead sheet is still available as are the craft skills for historic lead sheet roofing repair. (Radcliffe Camera, Oxford)

the reduction rolling mill, milled lead sheet was introduced and this has gradually taken over the building market from the traditional cast lead sheet.

There is no British standard for sand cast lead sheet, although its use is widely accepted in conservation work.

Milled lead sheet

Most of the lead sheet now used in building is manufactured on rolling mills and is described as milled lead sheet. It is manufactured to British Standard 1178: 1982 *Milled Lead Sheet for Building Purposes*. A standard range of thicknesses of milled lead sheet is shown in Table 5.1.

Table 5.1
Milled lead sheet − Standard range of thicknesses

BS 1178: 1982 Code No	Thickness (mm)	Weight (kg/m^2)	Corresponding thickness		Weight (lb/sq ft)
			Decimal (mm)	Nearest 1/64″	
3	1.32	14.97	0.052	3/64 +	3.07
4	1.80	20.41	0.071	5/64 −	4.19
5	2.24	25.40	0.088	3/32 −	5.21
6	2.65	30.05	0.104	7/64 −	6.17
7	3.15	35.72	0.124	1/8 −	7.33
8	3.55	40.26	0.140	9/64 −	8.26

(Based on reference 5)

Lead sheet and strip may carry colour markings for easy recognition of the thickness in store or on site, as follows:

Code No. 3 = Green Code No. 6 = Black
4 = Blue 7 = White
5 = Red 8 = Orange

Milled lead sheet is supplied by the manufacturer cut to dimensions as required, or as large sheets 2.40 m (nominal 8 ft) wide and up to 12 m (nominal 40 ft) in length. Lead strip is supplied in coils in widths from 75 mm up to 600 mm.

Direct manufacture (DM) lead/cast lead

Direct manufacture (DM) lead, also referred to as copperized cast lead sheet, is produced by a different method from milled lead sheet. In this process an internally cooled steel roll is rotated on the surface of a bath of molten lead, the lead solidifying on the roll to produce a continuous sheet. DM lead is very similar in appearance to milled lead. The product has an Agrément Board Certificate. A research programme at Cambridge University set up to investigate the weathering properties and performance of DM lead was completed in early 1988. The preface to BS 1178: 1982 includes a commitment to reconsider inclusion of this material in the standard in the light of the results of this study. As a result of the positive findings of the Cambridge study, this will now take place.

5.3 PROPERTIES AND DETERIORATION OF LEAD

Weathering of external leadwork

Lead is extremely resistant to corrosion by the atmosphere whether in urban, country or coastal areas. In time, lead develops a strongly adhering and highly insoluble patina of lead sulphate, which is silver grey in colour. Lead weathers in two stages. In the first stages of exposure it forms a surface film of lead carbonate which imparts a light grey appearance, is only loosely adherent to the lead and can wash off. Eventually the permanent patina of lead sulphate will develop as a result of reaction with carbon dioxide and, more important, with sulphur dioxide in the atmosphere to give an even coloured and adherent patina. The underside of lead sheet which is not exposed to the atmosphere does not develop this protective patina of lead sulphate and hence is susceptible to condensation corrosion.

Common causes of failure

Lead sheet has a high coefficient of expansion and this must always be taken into account when fixing. Many failures have arisen from the neglect of this principle. Many such failures have been attributed to the material itself, which has not been the case.

Common causes of failure of leadwork are:

- The use of sheets which are disproportionately large for their thickness (over-sizing)
- The use of an excessive number of fixings (overfixing)
- Failure of fixings
- Acid run-off from lichen-covered roofs
- Contact with incompatible materials
- Previous faulty repairs
- Condensation under the sheets.

Where failure occurs in lead sheet on an existing building due to bad design or workmanship it is most likely to be caused by fatigue, oversizing of sheets or the restriction of free movement by an incorrect method of fixing. Failure can also result from lack of or incorrect fixings.

Fatigue

Lead sheet fixed externally is subjected to conditions of changing temperature. Lead's relatively high coefficient of linear expansion coupled with a daily variation of temperature in the vicinity of, say, 40°C (104°F) means that the material experiences considerable dimensional change. Of course, the daily variation can be considerably higher than this – the seasonal change (summer to winter) can be up to 100°C (212°F). The detailing of leadwork must allow the lead to expand and contract freely, otherwise a ridge will form. Excessive fatigue stressing at the top of the ridge will eventually cause the sheet to crack. This is the most common cause of failure in lead sheet roofing and is due to oversizing (the use of sheets which are too large for their thickness) and/or overfixing (restricting the freedom of movement of individual sheets by too many fixings).

It is therefore of first importance to limit the size of each piece so that the amount of thermal movement is not excessive and also to ensure that there are no undue restrictions on this movement. Long experience has shown that it is quite practical to make provision for the thermal movement of lead.

Lack of fixings and failure of fixings

The lack of fixings and the failure of fixings are very common causes of the failure of leadwork. Slipping and buckling, which can occur as the result of these, are often misinterpreted as creep of the actual metal. Lead sheets which have slipped can usually be pushed back into place and fixed properly.

Creep

Creep is the tendency of metals to stretch slowly in the course of time due to their own weight. While lead is susceptible to this phenomenon, creep alone is rarely a cause of failure of lead sheet roofing. Creep failure can only result from considerable oversizing of sheets and would usually be preceded by a fatigue failure.

Condensation corrosion

Entrapped condensation on the underside of a lead covering can cause significant corrosion by slowly converting metallic lead to lead carbonate, lead oxide and lead hydroxide, which results in an off-white/pink/brown flaky powder which is not effective in the long term protection of the lead. The corrosion continues as an electrochemical reaction and can eventually break through from beneath. The effect is often referred to as 'sugaring'. This type of corrosion can be experienced in historic buildings which have undergone a change of use which involves the introduction of central heating accompanied by a decrease in ventilation or a large increase in the number of visitors. Such actions inside a building produce warm moist air which will rise to the underside of the lead sheet. If the sheet is at or below the dew point temperature of the air, the moisture vapour will condense on this lead surface which has not been exposed to free air and has not formed a natural protective patina. A never ending series of condensation/evaporation cycles will occur and the lead will continue to corrode unless the moisture is removed. This phenomenon has highlighted the value of the ventilation of lead sheeting which existed in traditional roofing construction.

During 1986–87 the Lead Development Association and others were actively involved in determining the causes of condensation corrosion under lead. Newsletter No 9 to the Lead Contractors Association (16 October 1986) considered condensation corrosion in relation to standard types of construction of lead sheet roofs. Concerning warm deck construction, where a vapour control layer and insulation is located above the roof deck and beneath the lead sheet, investigations and testing showed that 'in very hot summer conditions it is possible to produce sub-atmospheric pressure between the vapour control layer and the roof covering when the very hot metal is suddenly cooled by a heavy shower'. This made it possible for moisture to be drawn through joints and cause condensation corrosion on the underside of the lead throughout the roof space. This phenomenon is called 'thermal pumping'; it is known to occur where a completely effective vapour control layer is present. To overcome this possible problem a ventilated warm roof construction was exhaustively studied and put forward by the LDA, as the alternative. While this situation does not involve a traditional form of construction it has implications regarding modifications which may be proposed for a traditional lead roof.

Effective ventilation of the underside of lead sheet and hence the removal of moisture present is now, and will continue to be, a significant part of the solution of the condensation corrosion problem.

Run-off from lichens and mosses

Heavy dew or light rain which picks up organic acid from lichen or mosses growing on a roof and then drips slowly into a lead lined gutter dissolves its normal protective patina. Repetitive dissolving and reforming of the patina results in grooves being cut, although in the case of lead it will be many years before the gutter lining is penetrated. Where water drips off a roof, holes can be formed. Remedies to this problem include the removal of the lichens and mosses. A biocide may be used (see Volume 1, Chapter 2, 'The Control of Organic Growth'). An alternative method of preventing growth is by the insertion of copper strips at every ten courses of slates or tiles. Moisture running off these strips will carry coppper salts which inhibit the formation of lichens or mosses. Where it may be desirable to retain lichen or moss growths, or frequent maintenance is impracticable, a 'sacrificial' flashing can be provided at the drip-off points. These flashings are strips of lead sheet, inserted over the actual lead gutter linings, which are gradually allowed to corrode away in preference to the main linings. After thirty to forty years these sacrificial flashings can be renewed at relatively little cost compared with the renewal of the actual linings. (Copper, zinc or aluminium should not be used as a sacrificial flashing, as the acid run-off may corrode them through in only eight years.) The practice of applying bitumen paint along the drip-off line is not recommended. Apart from the need for regular painting, the acidity of the run-off is not neutralized and the corrosion problem is usually transferred to the area in front of the bitumen painted strip.

Attack from the free lime in Portland cement

Concretes and mortars based on Portland cement contain some free lime that can initiate a slow corrosive attack on lead in the presence of moisture. Any lead, such as a damp proof course, which is built in or laid within a bed of mortar should first be treated with bitumen. This does not apply to the 25–50 mm (1″–2″) turn in of flashings, where there is rapid carbonation of lime.

Corrosion resistance of lead

While lead has a very high resistance to corrosion there are, nevertheless, situations where precautions need to be taken.

Metals

The general experience is that lead can normally be used in contact with copper, zinc, aluminium and iron without corrosion by electrolysis being stimulated. In some marine and some industrial atmospheres it may be advisable to avoid metal-to-metal contact between lead and aluminium.

Acids

Lead resists corrosion by many acids including chromic, sulphuric, sulphurous and phosphoric. However, it is corroded by hydrochloric and hydrofluoric (common masonry cleaning agents), acetic (fumes of breweries and sawmills), formic (ants and other insects), and nitric acids.

Timbers

Dilute solutions of organic acid leached from hardwoods can cause lead to be slowly corroded. Also, the corrosive effect of continuous condensation on the inner face of roofing can be exacerbated when it takes up organic acid from hardwood members of the substructure.

There are organic acids in cedar roofing shingles which can be taken up by light rain and dew to form a dilute acid solution which will slowly corrode lead flashings onto which it runs. In this situation it is advisable to protect the lead with a coating of bitumen paint for a few years, during which natural weathering of the shingles will leach out the free acid.

5.4   GOOD LEADWORK PRACTICE

Prevention of most causes of failure involves ensuring that good leadwork practice is undertaken (see reference 5).

Firstly, the correct sizing and thickness of sheet should be used. Tables 5.2–5.5 are provided for reference. Compliance with the table may involve reconstruction of the substructure. The substructure should be inspected for damage during such work.

Table 5.2
Flat roofs — spacing of joints

BS 1178 Code No	Joints with fall mm	Joints across fall mm
4	500	1,500
5	600	2,000
6	675	2,250
7	675	2,500
8	750	3,000

(reference 5)

101

Table 5.3
Plain pitched roofs — spacing of joints

BS 1178 Code No	Joints with fall mm	Joints across fall mm
Up to 60° pitch		
4	500	1,500
5	600	2,000
6	675	2,250
7	675	2,400
8	750	2,500
Above 60° pitch		
4	500	1,500
5	600	2,000
6	675	2,250
7	675	2,250
8	750	2,250

(reference 5)

Table 5.4
Wall cladding — spacing of joints

BS 1178 Code No.	Vertical joints mm	Horizontal joints or height of panel mm
4	500	1,500
5	600	2,000
6	600	2,250

(reference 5)

Table 5.5
Gutter linings

BS 1178 Code No.	Maximum between drips (mm)	Maximum overall girth (mm)
4	1500	750
5	2000	800
6	2250	850
7	2500	900
8	3000	1000

(reference 5)

The second principle of good leadwork practice is that the *correct jointing and fixing techniques should be applied*. Fixings, while not restricting thermal movement, must be adequate to support the lead and, depending upon the degree of exposure, retain it in position. Joints must allow for thermal movement and yet still be weathertight for the position in which they are used.

5.5 REPAIR OF LEAD

Generally, a lead roof should be repaired rather than replaced unless the lead has failed or corroded on a large scale. Before any lead sheet is repaired or replaced, the cause of failure must first be understood and eliminated where possible.

If the deterioration is localized from either cracking or corrosion of the upper surface, for instance, the damaged section can be removed and a new section inserted by 'lead burning' (localized melting in an oxyacetylene flame) or welding. Solder should never be used to repair lead; its different thermal expansion will eventually cause it to break away. (Solder is comprised of 2 parts lead: 1 part tin. A soldered joint is different in appearance from a lead-burnt joint; it is usually wider and has a smooth, wiped surface instead of the series of small lumps of the lead-burnt joint.)

Where a ridge of lead has formed in a sheet it is not possible to drive the lead back into place. The lead in the ridge zone adopts a crystalline structure which does not accept welding. Corroded lead cannot be welded either. A section 50 mm (2 ins) wide to all sides of the failure must be cut out, the area properly cleaned and a new patch which overlaps the hole by at least 10 mm (½ in) lap welded on. A butt joint between the existing and repair lead must be avoided to prevent the risk of fire.

Where lead is severely deteriorated from erosion, a first aid repair in the form of an adhesive tape repair bandage can be used. Mastics, asphaltic or bituminous roofing compounds should not be used because they obscure the source of the problem and will significantly reduce the re-sale price of the old lead.

The use of patination oil

The initial white patina of basic lead carbonate can be washed down a roof, causing streaking and staining of the lead and any adjacent materials. The formation of the initial patina can be controlled by an application of a smear coat of patination oil on the lead as soon as is practical after fixing, and the staining problem avoided. The oil should be applied with a mutton cloth and care must be taken to ensure the patination oil is applied thinly and evenly, or the application lines will show.

Cleaning lead

It is rare that lead roofs need to be cleaned. However, they should be kept clear of corrosive pigeon droppings. Dirt on exposed lead should be removed with a neutral pH soap in warm water applied with a bristle brush. A lead cleaning gel is available from lead manufacturers. Basic lead carbonate can be removed with a number of EDTA-related products (EDTA = ethylene diamine tetracetic acid). Microblasting with glass beads or coconut shells, which is used to clean lead sculptures, has severe health and safety implications.

5.6 LEAD FLASHING DETAILS

The purpose of a lead flashing at an abutment is to weather the junction between

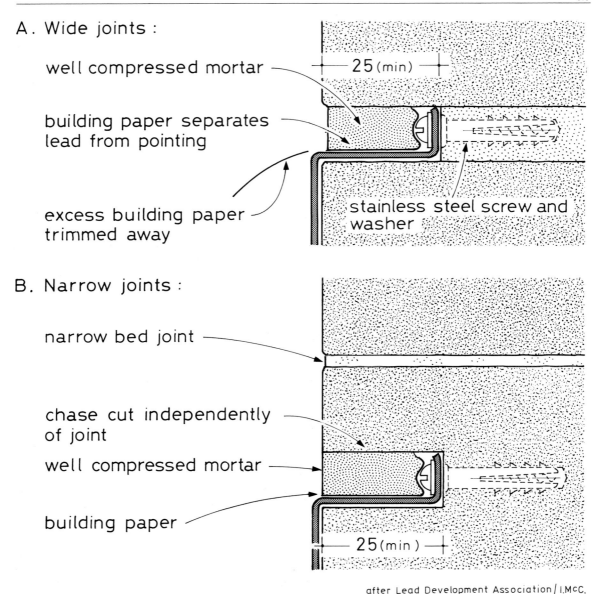

A. Wide joints :

well compressed mortar

25 (min)

building paper separates
lead from pointing

excess building paper
trimmed away

stainless steel screw and
washer

B. Narrow joints :

narrow bed joint

chase cut independently
of joint

well compressed mortar

building paper

25(min)

after Lead Development Association / I.McC.

Figure 5.1   Recommended details for the fixing of lead flashings at abutments
*If properly executed, these details allow the lead flashing and the pointing
mortar to move independently and thereby retain a weatherproof joint.*

a wall and an adjacent surface as, for example, a parapet gutter and parapet wall, a pitched roof and a flue terminal. This common detail is so frequently executed incorrectly and can cause so many costly problems associated with water penetration that the Lead Development Association has made specific recommendations of modifying the standard detail in all but the most sheltered situations. Figure 5.1 illustrates recommended procedures for flashing into narrow joints and wide joints in historic building fabric.

5.7 THE USE OF LEAD SHEET SUBSTITUTES

When is a lead sheet substitute acceptable and necessary?

There is an increasing interest in various substitute materials for replacing lead sheet which has deteriorated or has been stolen from roofs, often incurring considerable incidental damage. The choice of substitute materials is, of course, at variance with the general conservation policy of repairing or replacing like with like; but the primary need to keep roofs watertight and, in some areas, the high incidence of theft sometimes justify the consideration and use of alternatives. The relatively high cost of replacing a lead roof with new lead, especially on, for example, a small church with minimal funds, will also be an influencing factor.

In most circumstances the ideal decision will be to replace lead with lead, new or re-cast, and it is the first material which should be considered. Lead is the original and traditional sheet metal roofing material and the most acceptable in visual terms, and its behaviour is so well known that correctly designed and detailed lead coverings should remain trouble-free and durable for a very long period.

Choosing a substitute material

A substitute material for lead sheet roofing should not compromise the desirable qualities and properties of lead which contributed to the long term appearance and wellbeing of the structure beneath.

Where the case against replacement with lead is considered strong enough, aluminium, zinc, stainless steel, terne-coated stainless steel and lead-clad steel have all been suggested as substitutes at one time or another. Not all these materials are considered satisfactory substitutes for lead sheet roofing. A description of their properties, performance and suitability follows.

Where lead cannot be afforded it is better, at least, to have a visually similar material of the same sheet size rather than go to any totally alien material; substitutes should have similar weathering characteristics to lead. Where lead is the covering on steeply pitched roofs or domes on buildings of considerable architectural importance it is, of course, unlikely that anything other than lead will do.

Where the decision has been made to replace a lead roof covering, the choice of material should take into account the following criteria:

- The character of the existing roof; its colour, finish, joint type and profile.
- The likely performance of the substitute material.

● The cost of re-covering in conjunction with the long term value of the replacement material.

Caution is advisable in taking the word 'substitute' too literally. Modifications will need to be made to accommodate the new material. No substitute has the versatility of lead and can always be used in the same way.

The trade

All the lead sheet substitutes listed above require hard sheet metal techniques. Traditional roofing contractors may not like using them and may be unfamiliar with their working and fixing. The steels are more difficult to cut, handle and work than other metals and their cut edges are dangerous to workmen. A contractor experienced in the substitute material in question should be selected.

The harder sheet substitutes may also be difficult to use on irregularly shaped buildings which may need the formability of lead. In addition, while a few may produce a similar appearance to a traditional lead roof, their fixing involves a different detailing.

Costs

Assumptions on costs should give way to proper cost comparisons between lead and the substitute under consideration, as well as between substitutes.

In approximate order of increasing cost the possible substitutes may be ranked as shown below, but of course the sheet thickness and the system of jointing (batten rolls or standing seams) will make considerable differences and must be taken into consideration:

> Aluminium
> Zinc
> Stainless steel
> Terne-coated stainless steel
> Copper
> Lead/lead-clad steel

While lead may be up to 30 per cent more expensive than most substitutes selected to give a similar lifetime, detailed cost analysis may show the margin to be smaller than originally thought due to the design of the roof in question.

The installation cost of a substitute material is influenced considerably by the complexity and shape of the roof in question. Some roofs may include complex details which can only be carried out satisfactorily in lead; in some guttering situations the lead has not performed well and a substitute material may need to replace it.

The capital cost plus any maintenance cost should be carefully evaluated against the durability of the material selected. Substitute materials to lead should have a comparable life expectancy and certainly a similar cost per year of useful life.

Theft

The formability of lead sheet roofing makes it easy to remove. Early consideration should be given to deterring vandalism by means such as floodlighting. The deterrent of extra fixings is not recommended since these frequently have the effect of restricting free thermal movement. Where lead coverings are particularly vulnerable to theft, renewal in lead may not be insurable and materials with a low scrap value or which are difficult to strip may need to be considered.

Properties of substitute metal coverings

Table 1.1 compares the approximate weight of lead roofing and the substitute sheets.

The substitute materials are substantially lighter than lead. When lead is replaced with an alternative metal it is important to consider the method of fixing, especially at free edges, to ensure that the substitute material is not damaged in high wind conditions.

Noise is an additional factor for consideration when a lead sheet substitute is to be used. Although all substitute metals, apart from zinc, have lower coefficients of thermal expansion than lead, some hard metals will 'creak' unless the fixings are so detailed that thermal movment is not impeded (see Table 1.1).

The impact of heavy rain on hard materials such as stainless steel without insulation underneath can create problems inside a building where noise is a critical factor.

Most hard sheet metals will not lie as evenly as lead, giving the roof surface a quilted appearance. Workmanship and joint design modification such as the use of solid rolls can only minimize this.

Where a substitute material is to be used, replacement and maintenance work should be organized so that at least a whole pitch or bay of a roof is completed at one time.

An evaluation of substitute materials

Precautions must be taken to ensure that there is no risk of bimetallic corrosion wherever different metals are used together (see Table 1.2).

Copper

Copper is sometimes suggested as a substitute when total replacement of lead is considered, but the new and weathered colour of copper make it a visually unacceptable alternative to lead sheet. Furthermore, if there is a water run-off on to the roof from lichen-covered masonry or tiles there is a risk of corrosion. See Chapter 4, 'Traditional Copper Roofing'.

Aluminium

The brightness of new aluminium sheet takes some time to dull down and even after 10−15 years it will only superficially resemble lead. Generally, aluminium would not be considered an acceptable substitute to lead, especially in highly polluted environments. If there is a water run-off onto the roof from lichen-covered masonry or tiles there is a risk of corrosion.

Zinc

Zinc was used as a cheaper substitute for lead roofing after the Industrial

Revolution. This practice is not general today. If there is a water run-off onto the roof from lichen, copper or timber shingles there is a risk of corrosion. Zinc is much less malleable than lead and requires hard metal details which lead does not. Zinc sheets are significantly thinner than lead and overall it is considerably less cost effective. See Chapter 7, 'The Repair and Maintenance of Traditional Zinc Roofing'.

Lead-clad steel

Lead-clad steel consists of 0.75 mm lead sheet cold roll bonded to one side of 0.1 mm terne-coated steel sheet (not stainless steel). It has the advantage of a low theft risk coupled with the appearance of lead. However, it also has the significant disadvantages that panels need to be pre-formed off site, so that extreme accuracy in measuring is required, and that any raw steel exposed at cut edges or by scratching must be completely protected by solder to prevent corrosion. This must be balanced against the advantages of reducing on-site time. Although the surface of the steel is protected by terne-coating, the back of the sheets are not, and should be given a paint coating for additional protection. Lead-clad steel may be more expensive than lead.

Stainless steel

Stainless steel is a corrosion resistant material which has been used as a roof covering since the late 1960s. Type 304 is suitable in normal situations, while type 316 is preferable in more aggressive industrial and marine environments. Both grades can be obtained with a low reflective finish which is substantially less reflective than ordinary stainless steel. However, this finish is more reflective than lead and does not alter significantly on weathering. Stainless steel is consequently not a good visual match for lead. Traditional drips are difficult to form in stainless steel and the gutter detail may need to be modified or lead may need to be used in gutters to form these. Vertical joints can be formed using either the batten roll or standing seam detail. Stainless steel strip normally employed is very thin and razor-sharp, so that operatives must wear protective gloves.

Stainless steel does not always lie happily against uneven surfaces. For the more complicated details it may still be advisable to use lead for flashings and gutter linings. In gutter situations where the run is relatively straight and there was an original problem of lack of fall, resulting in an inadequate number of drips in the lead lining, stainless steel can provide a successful replacement.

Terne-coated stainless steel

This material is stainless steel which has been coated with lead–tin alloy. It combines the corrosion resistance of stainless steel with the added protection of the lead–tin coating. The design and detailing of the material is as described for conventional stainless steel. The terne-coating of this material quickly loses its initial reflectivity and then weathers to a 'lead-like' appearance, slightly lighter than weathered lead. The darkness of its final colour depends on atmospheric conditions. The result is probably one of the best visual substitutes for lead roofs. It does, however, require different detailing from lead sheet.

Where the situation is not straightforward, lead flashings and copings may need to be used. The material can be used in conjunction with lead sheet roofing where problems are experienced in replacing a lead-lined gutter.

Terne-coated stainless steel is available in slightly varying compositions.

1 A Type 304 stainless steel coated with an alloy of 80 per cent lead to 20 per cent tin. The coating is 30 microns thick and is metallurgically bonded to the stainless steel.
2 Type 304 and Type 316 stainless steels are available coated with an 85 per cent lead to 15 per cent tin alloy, 2–3 microns thick.
3 The third type of terne-coated stainless steel is comprised of a thin terne-coating of 90 per cent lead to 10 per cent tin applied to Type 430 (ferritic) stainless steel.

Concluding comments

The selection of a substitute for sheet lead should not be made lightly.

The appearance of lead is matched most closely by the terne-coated materials. It is important that the terne-coating should be substantial enough to continue to provide this appearance and the associated protection for a period of time comparable to the lifetime of the original lead. Terne coating on stainless steel gives significant long-term performance which begins to approach that of the original lead. The detailing of the substitute sheet must be altered from lead detailing, especially on low pitches, and contractors experienced in hard sheet metal fixing must be employed for the work. A sheet width comparable to the original is often desirable architecturally and this will also be a factor in substitute selection. Complex details and flashings will still need to be lead. Terne-coated stainless steel of appropriate quality may be a useful substitute for failed lead gutters with inadequate falls even where lead is retained for the roof cladding.

REFERENCES AND FURTHER INFORMATION

References

1 British Standards Institution, BS 1178: 1982, *Milled Lead Sheet and Strip for Building Purposes*.
2 British Standards Institution, CP 143 Part 2, *Draft Specification for the Design and Construction of Fully Supported Lead Sheet Roof and Wall Coverings* (Draft out for public comment).
3 British Standards Institution, BS 6229: 1982, *Flat Roofs with Continuously Supported Coverings*.
4 Ecclesiastical Architects and Surveyors Association, *Corrosion of Lead Roofing*, in preparation (1988).
5 Lead Development Association (LDA), *Lead Sheet in Building: A Guide to Good Practice*, LDA, London, 1978 (revised 1984).
6 Lead Development Association: *Leadwork* quarterly journal since September 1979, LDA, London.
7 Lead Development Association, *Lead Sheet Flashings for Slate and Tile Roofing*, LDA, London, March 1981.

8   Melville, Rodney, 'Non-traditional Material in Church Roofing', *Conference on New Materials in the Conservation of Churches*, Council for the Care of Churches, November 1981, pp 8–17.

9   'Sheet Roofing Materials – Products in Practice' *AJ* Supplement, *The Architects Journal*, 28 March 1984, pp 23–39.

See also the Technical Bibliography, Volume 5.

Further information on lead roofing

Lead Development Association
34 Berkeley Square
London W1X 6AJ
Tel: (01) 499 8422

6 LEAD SCULPTURE AND ITS CONSERVATION*

Lead has been used ornamentally for several centuries principally in the form of church fonts and figures, urns and fountains for gardens, civic squares and stately homes. The repair, maintenance and conservation of these items requires a special balance between the skills of the plumber/craftsman and the sculptor/artist and should therefore remain in the realm of the metals conservator. This chapter deals mainly with lead sculpture, its manufacture and decay, and reviews the specialist skills required for its repair and conservation.

6.1 THE MANUFACTURE OF LEAD SCULPTURE

Lead began to be used for ornaments and statues in Britain in the fifteenth century. Early small figures were cast solid. Up to the end of the eighteenth century the two main casting methods were the 'lost wax' method and 'conventional (clay mould) casting'. From about 1790 through to 1920 the 'Victorian slush cast' method was used. Conventional casting and Victorian slush casting are the predominant methods used in forming lead sculpture.

In *conventional casting* the design was first produced in clay, usually in sections. A core and a mould were then produced in a sand/lime mixture, dried and baked. The core and outer mould were located and held apart by iron chaplets (pins) and the hot lead poured in between. The sections of the casting were reunited and finished on the surfaces. Some large castings were supported with wrought iron armatures. These were often in themselves structurally incomplete and relied on bedding into the core for additional support.

The *Victorian slush casting* system did not use a mould with a core inside. Instead, molten lead which had partly cooled (hence the term 'slush') was poured into an open mould and left for a few minutes until it had chilled against the cold wall of the mould, the excess being poured out. It was difficult and rare to achieve an even thickness of casting by this method. Usually the first attempt was too thin, especially at the edges and corners. The thickness was then built up by a sequence

* RTAS wishes to acknowledge the contributions of Naylor Conservation, Telford.

111

of pours which often produced a series of incontinuous foils of lead. Dirt and corrosion products are often found in between these. These thin laminations make it impossible to lead burn sections of a lot of nineteenth-century sculpture, as the dirt and corrosion products act as an insulant.

The sections of Victorian casts were often soldered together, which produced a flimsy joint. As no armature was inserted the structural integrity of the sculptures relied on the joints. The combined result was a weak and unstable casting in a structurally deficient form which is difficult to repair.

Once the casting was complete, finer details of the design were carved in, using wood carving tools. A surface finish was then applied. It is rare that lead sculptures were ever left to look like lead, particularly before the nineteenth century. Lead was often used as a cheap imitation of sculpture in stone. The sculptures were rendered in natural colours, made to look like stones such as Bath stone or marble. They were at times gilded and bronzed (reference 1). During the late nineteenth century it was fashionable for lead sculptures to be made to look antique. A wide variety of processes and materials, including hydrochloric acid, lime putty, tea leaves, scorching and thin oil paint followed by copperas, were used to achieve this (reference 5, p 182).

6.2   ASSESSING LEAD SCULPTURE

Poor design or manufacture, physical change or chemical attack, physical damage, inappropriate repairs and maintenance, are all factors which contribute to the deterioration of outdoor lead sculpture. Therefore, the inspection, assessment and subsequent repair, maintenance and conservation of lead sculpture is not the sphere of a lead plumber, but rather the specialist metal conservator. The conservator should be engaged as early as possible to enable the schedule of the project to be realistic and proposed works appropriate to the condition and historical value of the sculpture.

Problems of design and manufacture
The first category of problems which a lead sculpture may be experiencing originate in its design and manufacture.

- The design of the piece may have caused an over-stressing of the lead. Relatively heavy angels' wings can shear if the area of connection is too small.
- Lead cannot always accommodate the stresses of a sculpture, e.g. an angel poised on one leg. Forms which are thin and light, e.g. laurel leaves, are easily stolen.

During the manufacturing stage problems relating to casting, fixing and internal support may be built in.

- The cast may be honeycombed and porous where water vapour from a damp core has bubbled through.

- Extraneous material such as dross may have been included in the pour.
- Molten lead which has been in contact with a cold mould may have solidified prematurely into laminations which provide discontinuous, weak points.
- Parts of the casting may be excessively thin and therefore more vulnerable to deformation or rupture. The vulnerability extends to the thinner areas of a casting which has excessive variation in thickness. Thinness may also be created after casting when the sculpture is carved.
- Iron chaplets, the locating pins between the two parts of the mould, which have not been removed are most likely corroding.
- Foundry joints between the sections of the casting may have been poorly done. A good join is a homogeneously welded join, and this only occurred if the pour metal was hot enough. Often mechanical joins were made into which molten lead was poured, and often the lead was not hot enough, resulting in a joint which can easily be pulled or jacked apart. Poor quality joins may also be found where the design of the sculpture made access difficult.
- The original sand:lime core may have been subjected to rising damp and penetrating damp, encouraging corrosion of the inner surfaces of the lead by the retention of moisture at the surface.
- Internal ironwork of the original armature may be corroding.

Deterioration, weathering and previous repairs

As a lead sculpture endures its environment it can experience distortion, fracture, surface damage and deterioration of its surface finish.

- The original sand:lime core may have been replaced with Portland cement. Alkali salts in the Portland cement can react with and oxidize the lead. The resultant products provide a form of corrosion jacking. Cement cores are usually found to be fractured and of limited effectiveness.
- Lead sculptures which have begun to collapse for one reason or another may have received inappropriate supports, such as a steel rod fixed into the lead with screws. A concrete base may have been cast around the base of the structure in an attempt to provide support.
- Previous repairs are the most common form of damage to exterior lead sculptures. Unsympathetic repairs using solder, putty, mortar and filler may have been carried out *in situ*. Good clues as to the original surface may remain under those repairs. Alternatively, a relatively minor damage may have been pushed in to provide a key for the filler mix.

6.3   UNDERTAKING REPAIR AND CONSERVATION

It bears repeating that any work on a lead sculpture should involve a metals conservator at all stages from preliminary survey and recording through to deterioration and environmental assessment and the scheduling and undertaking of repairs.

The repairs procedure will usually involve transport of the sculpture to a workshop where some or all of the following works may be undertaken: taking and analysing paint samples, cleaning to remove corrosion products and deteriorated paint surfaces, removal of previous repairs, dismantling and removal of armature and core, reshaping of distorted areas, repairs such as patching, lead burning and reattachment, provision of a new stainless steel armature and repainting.

Some conservators also fill repaired lead sculptures with closed cell polyurethane foam to ensure all surfaces receive sufficient support. While this practice is generally sound, in the longer term there is a slight possibility that organic acids which develop as breakdown products of the foam will corrode the lead. Should further work on the sculpture be required in the future the foam will be very difficult to extract mechanically and problems will arise if further lead burning is required. However, it is often the case that the advantages of filling a lead sculpture with foam outweigh the potential disadvantages.

The term 'leadburning' refers to the welding of lead. The filler material used is lead, not solder. Leadburning is a skilled operation which requires experience in controlling the flame of the blowpipe. (For further information see Appendix B, reference 2.)

It has already been mentioned that the Victorian slush cast method often produced a weak and unstable casting. It should be noted that castings of this sort are consequently difficult to repair. This sort may not be resilient to outdoor conditions either before or after treatment.

6.4 MAINTENANCE OF LEAD SCULPTURES

It is best to avoid *in situ* repairs of lead sculpture. The orientation of the sculpture surface is usually not conducive to good workmanship. Furthermore, there is always a strong chance that water will be sealed in.

Propping of a leaning sculpture can be dangerous as the stress regime may simply be altered rather than removed. Screwing into lead and strapping back is ineffectual and to be avoided completely. Tying back with wire is quickly damaging.

If a sculpture is unstable it will probably be better to take it down. Removing lead sculptures requires experienced knowledge of the behaviour of such objects. The right sort of slings such as wide canvas webbing must be used — chains, wire ropes, hard ropes and welding straps are to be avoided at all cost. The slings must be placed correctly. Lead is very malleable and very heavy. A lot of damage can be done if the sculpture is not supported in the right places. The most fragile, thin areas such as ankles and necks will be the first to bend and fracture if proper support is not given. Crushing, tearing and distortion can be expected if a lead sculpture is lifted incorrectly.

Once a sculpture is down it must be supported correctly in a dry area. This can be achieved using polystyrene blocks with plenty of bearing area.

Any cleaning of lead such as the removal of bird droppings should be done with soft cloths and gentle pressure. All types of wire brush must be avoided. Even natural bristle brushes used with pressure can score the surface of lead.

6.5 # CASE STUDY: THE REPAIR AND CONSERVATION OF A LEAD SCULPTURE

Through the Research, Technical and Advisory Services, the Ornamental Ironsmiths' Workshop and the Stone and Wood Carvers' Studio of English Heritage undertake repair, conservation and replacement work on monuments and objects in the care of English Heritage. On certain occasions and when the work programmes permit it, work is also undertaken on projects for private clients. For the Ornamental Ironsmiths such a project took place in early 1987. It involved the lead sculpture 'Charity' from the Fishermen's Hostel, Great Yarmouth, and formed part of the repair and rehabilitation work that had taken place there.

While the work was based at the Ironsmiths' workshop it involved other specialist craft analytical services with English Heritage which combined to provide the multidisciplinary approach that the conservation of such objects demands. The project was co-ordinated by the Research, Technical and Advisory Services.

Survey and condition

The sculpture was comprised of the maternal figure of Charity with an infant in her right arm and a young child clinging to her left knee (see illustration p.116). It was located in the forecourt of the Fishermen's Hostel, which was built in 1702, and appeared to date from that time, although the present position may not be correct. The sculpture was a very finely cast and modelled piece of work. It had been declared outstanding by the English Heritage Inspectorate, and was included in *English Leadwork, Its Art and History* by Lawrence Weaver (1909).

The sculpture was first surveyed by RTAS in May 1986. The figures appeared generally battered and neglected. The surfaces had been coated with a black bituminous paint which had degraded to reveal other light decorations below, generally obscuring the original visual intention of the design. Charity's right arm was detached and had been kept in storage along with the infant it once held. The upper arm had been capped with a lead stump at the elbow. Charity was leaning forward, and distortion and fracturing in her right foot showed the significant extent of previous movements of this kind. A disfiguring mild steel brace had been fixed to the base of her spine in an attempt to arrest further movement. A second such brace was fixed under the left elbow of the young child at her knee. A concrete plinth had been cast around the feet of the figures in a misguided attempt to hide any distortion which had occurred and to provide stability.

Closer inspection of the surfaces revealed inappropriate previous repairs in solder, several large dents possibly from a fall, and an extensive number of marks and graffiti which were attributed to vandalism. Several of the foundry joints between cast sections had begun to fracture. As is usual with this sort of work, the true condition became clear once the figures had been removed to the workshop and cleaned.

Works undertaken

RTAS took charge of the figures in January 1987. Before any work commenced,

115

In Britain, lead was cast for statues from the fifteenth century, but most prolifically during the seventeenth, eighteenth and nineteenth centuries. Particularly in the seventeenth and eighteenth centuries, the statues were painted either naturalistically or to imitate natural stone. The deterioration of a lead sculpture can involve corroded wrought iron armature, deformation due to incorrect support, impact or design problems, fracturing of foundry joints, defects in the casting and the damage due to inappropriate repairs such as the use of solder. The repair and conservation of lead sculpture is the domain of the sculpture conservator. It should be preceded by research including determination of the original paint colour if any. ('Charity', Fishermen's Hospital, Great Yarmouth)

paint samples were taken for microscopic analysis to determine the original colour or colours. Lead figures were commonly painted, often to imitate stone and at times in naturalistic colours. Close visual inspection had already revealed at least two colours.

The bituminous paint was removed by non-caustic gel. Wood implements such as modelling tools and orange sticks were used to encourage paint out of crevices. Corrosion products, particularly on the internal surfaces, were removed with a sequestering agent in gel form.

Inspection at this stage showed the figures to be without any core or armature. The castings were generally 4–6 mm (about ¼in) thick. This thickness had enabled the figures to stand as well as they had. In a few places the casting thickness went down to 2–3 mm (⅛in), and fractures had occurred in a few such areas.

Once the paint on Charity's torso had been removed it could be seen that the detail areas such as the breasts, fabric folds and the belt had been made up in a lead oxide in order to cover large areas of porosity where detail was missing. When the lead cap to Charity's right arm was removed, the arm was found to have been stuffed with newspaper (dating from December 1960) and then filled with foundry loam. The solder repair at this point was removed to reveal a large section of the original drapery.

Wherever possible, fractured joints or other tears were made good by lead burning. As this work proceeded it became clear that while the figures were artistically superlative, the quality of their casting was poor and had frequent and extensive dross inclusions. Whenever lead burning was made impossible because of this an inert epoxy-based filler had to be used.

Charity was fitted with a stainless steel armature which extended through the new stone cap on which she now stands to an adjustable fixing within the new brick plinth. An armature was also provided for the right arm, which had broken away due to an inherent design fault as well as thinness of casting. At the bearing areas the stainless steel was formed to the internal shape of the casting. It was then held 6–12 mm (¼–½in) away from this and a firm polyester rubber pad poured around to ensure there were no areas of point loading. Lateral stability was provided by two stabilizing bars in the lower hip area and locating dowels into the stone through original fixing holes at the edge of the drapery.

Charity's right arm was refixed by the lead burning. The original fixing points of the baby had been located during the cleaning process and he was refixed in his original position by a system of small stainless steel dowels.

The dented areas of the rear of the standing child were gently heated and returned to their original shape.

The lower areas of Charity and most of the standing child were peppered with dents up to 25 mm (1in) long and 3 mm (⅛in) wide and deep, probably due to vandals. These marks scarred this area of the work. The larger ones of these were made good with filler. While at no stage during work was it intended to make the group look anything other than an early eighteenth-century work of art, it was felt that this cosmetic work was necessary.

On completion of the works the figures were blown clean and washed down to ensure a thoroughly grease and dust free surface for the application of the paint. All paint layers were applied by brush. Following the application of two coats of traditional lead based primer paint, Charity was repainted in the four naturalistic colours which had been identified by paint analysis. The painting was undertaken by the skilled artists of the English Heritage Conservation Studio.

The figures were refixed on a new Portland stone cap. The main bar of Charity's armature was bolted to adjustable stainless steel fixings within the rebuilt plinth. The standing child was secured by means of stainless steel locating dowels in its feet into the stone and small stainless steel pins into the main figure.

On completion of the works the architect and owners were supplied with an illustrated report on the works undertaken and maintenance recommendations.

ADDENDUM: THE COLOURS OF LEAD COMPOUNDS

The typical colours of various lead compounds are:

Uncorroded lead	blue grey
PbO (litharge)	yellow to buff
Pb_3O_4 (minium/red lead)	orange red
$Pb(OH)_2 . 2PbCO_3$ (basic carbonate/white lead)	white
$PbCO_3$ (carbonate)	whitish grey
$PbSO_3$ (sulphite)	black
$PbSO_4$ (sulphate)	silvery grey

(Reference 4, page 29)

Melting point for lead = 327.4°C

The creep range occurs at room temperature. The purer the lead the more sensitive it is to creep.

REFERENCES

1   Jackson-Stops, Gervase, 'New Deities for Old Parterres – The Painting of Lead Statues' *Country Life*, 29 January 1987, pp 92–94.

2   Lead Development Association, *Lead Sheet in Building: A Guide to Good Practice*, LDA, London, 1978 (revised 1984).

3   Naylor, Andrew, 'Naylor Conservation – Recent Work of a Specialist Conservation Service', ICCROM Conference, *Conservation of Statuary and Architectural Decoration in Open-Air Exposure*, Paris, October 1986.

4   Simpson, Lorne Gordon, 'Conservation of Ornamental Leadwork', MA Conservation Studies, University of York, Institute of Advanced Architectural Studies, August 1986.

5   Weaver, Lawrence, *English Leadwork: Its Art and History*, London, B T Batsford, 1909.

See also technical bibliography, Volume 5.

7 THE REPAIR AND MAINTENANCE OF TRADITIONAL ZINC ROOFING*

Rolled zinc has been used as a roofing material for approximately 150 years. Whilst its most predominant usage has been on the Continent, zinc was used in Britain and it therefore, at times, forms part of the repertoire of historic roofing materials. This chapter looks at the history of zinc roofing in Britain, the properties of the material, traditional fixing methods and appropriate repair, conservation and maintenance techniques.

7.1 HISTORY OF ZINC ROOFING

The history of zinc is most ably covered in a series of three articles by Mr George H. Ridge, former technical adviser of the Zinc Development Association (see References), and these have been referred to extensively in this section.

Early zinc production and usage

In 1738 William Champion produced the first commercial quantities of pure metallic zinc at Warmley, and in 1740 a zinc works was established in Bristol. Hot rolling of sheet zinc was patented in 1805 by Hobson and Sylvester of Sheffield but their work was apparently not a commercial success, for nothing further was heard of them (reference 7, p.2). In 1811 the first commercial production of zinc sheet began in Liège, Belgium (reference 4, p.111). Following this, zinc sheet became the predominant roofing material in France, Belgium and Germany and today is still a very widely used roofing material on the Continent.

*RTAS wishes to acknowledge the assistance of Mr George H. Ridge in the preparation of this chapter.

119

Zinc roofing only became marginally as popular in England. The earliest reliable reference to zinc being used for roofing dates from 1832, when 23in (575 mm) square tiles were used on a factory in Bristol. In 1835 the cloisters of Canterbury Cathedral were roofed with zinc laid between round topped wood rolls, the rolls being covered with lead cappings (reference 6, p.2). Zinc roofing developed slowly but received a sudden impetus when it was used to cover the roofs of many of the new railway stations. Several small to medium-sized stations, mainly in rural areas, still retain their original zinc roofs, which are now about 100 years old. It is thought that zinc was used on the stations because it could be laid at low pitches, was light in weight and could compete in cost with slates and tiles, which required more substantial and complicated substructures. Lead was used on the main termini.

Zinc was used on roofs as tiles, ornamental pieces, and sheets fixed by the roll cap system or the standing seam system. All the roofs referred to above, except the Bristol one, were laid as sheet on the roll cap system and this remained the most usual way for laying zinc roofs in Britain. In contrast, Continental practice of the eighteenth and nineteenth centuries was to lay zinc on roofs of much steeper pitch and in the form of tiles. Zinc tiles did not appear in Britain until about 1860.

The corrugation process was patented in Britain in 1829, but the corrugated zinc produced was found to sag at normal temperatures. In 1837 the galvanizing process was patented in France and England. This began the use of zinc as a protective coating for ferrous metals. The first zinc oxide paint was made in France in 1781 and became commercially successful in about 1850. This paint and later zinc-chromate paint were found to be good inhibitors against rust on iron.

The standing seam system

In standing seam roofing, the joints running with the fall of the roof are formed by turning the sides of the zinc sheets up at right angles and then welting the edges to give a standing seam, fixing clips being incorporated in this seam. This produces the characteristic thin lines of the seam running down the fall. The zinc sheets are fully supported by the substructure. The standing seam system requires more labour and greater skill to do than either the roll cap or the Italianized systems.

The first recorded zinc roof (1810, Royal Iron Foundry, Berlin) was laid on the standing seam system using small sheets (reference 8, p.10). After this the standing seam system was virtually abandoned, but it was revived in the 1930s in Germany and has increased in popularity on the Continent since World War Two. The forming of the seams is now greatly assisted by machines. The system is used in Britain on new zinc roofing.

Zinc tiles

Zinc tiles were developed on the Continent around 1811. They were originally a plain diamond shape with folded edges which interlocked. The tiles were laid over boarding and secured with at least three nailed clips engaging with the folds. Tiles were also made in various sizes from 12ins (300 mm) to 23ins (575 mm) square, and also as equal sided parallelograms.

Use of plain zinc tiles declined with the development of the roll cap system (see following section) because this was cheaper and quicker to lay, but the use of ornamental embossed tiles expanded rapidly.

Ornamental tiles were mainly used on mansards, turrets and cupolas. Their large scale production in a wide range of patterns reached its peak in the period 1880–90 and declined gradually after that. The tiles were originally produced mainly in Belgium and France and exported in large tonnages, appearing in Britain in about 1860. Manufacture was set up in Britain by a number of firms. The most popular style was a shield-shaped tile with an embossed fleur-de-lys pattern. Blanking and forming operations were mostly carried out in hand-operated presses. Drop stamps were used more rarely for larger tiles or where particularly high relief ornamentation was required.

To cater for the mid-Victorian taste for high ornamentation, zinc sheet was pressed into highly decorated tiles, ridge capping and hollow ornaments such as gargoyles. Hollow components were made in two or more pieces with a drop stamp and soldered together. Ornamental capping and finials were not only used on zinc roofs but were also applied to the hips and ridges of slated and tiled roofs, as well as to dormers. Many types of ornamental zinc dormers were also produced during this time. Some required a timber support structure but most derived sufficient rigidity from the complex system of the mouldings.

Ornamental tiles were generally of fairly heavy gauge (15–16 SWG to 19–20 SWG) and, as they were usually laid at steep pitches, performed well even in more polluted industrial atmospheres. Their gradual disappearance was primarily due to changes in taste and quality of workmanship.

The roll cap system

The use of plain zinc tiles declined when larger zinc sheets, usually 3ft by 7ft or 8ft in Britain, became available and the roll cap system provided a system which was cheaper and quicker to lay. This followed the development of more powerful rolling mills, which increased output by pack rolling.

For the roll cap system the sheets were formed into shallow trays, originally on site, later by machine in the workshop. The side upturns were held by clips and the cross joins formed into drips or welts according to the pitch of the roof. The formed sheets were laid between wood-rolls which were subsequently covered with zinc capping fixed independently of the sheets. In early roll cap roofs the wood-rolls were pyramid shaped. Difficulties in fixing the capping meant these were soon superseded by round topped capping. Round topped capping required a saddle to be soldered at each top end and a purpose made stop end to be soldered at each lower end, so this in turn was superseded by square wood-rolls where these details were formed by simple folding operations.

Zinc roofing after the 1920s

The demand for zinc roofing was curtailed by the outbreak of World War One. Between the wars many building additons were roofed in zinc, sadly, often by workers with inadequate experience of zinc. The reputation of zinc roofing was further damaged by the use of a particularly thin gauge of zinc under the justification of cheapness and ease of working, and by the late 1920s zinc was generally considered to be suitable only for temporary work.

Italianized zinc roofing

This semi-self supporting system of zinc roofing was developed after World War Two for covering large industrial buildings such as warehouses. Factory-formed sheets of zinc with rafter-shaped ridges pressed into them were laid directly onto the rafters without the use of boarding or other structural checking. The sheets were overlapped and fixed to the tops of the rafters. The Italianized system was suitable for roofs of 22½ degree pitch or steeper, as well as for cladding vertical surfaces.

Current zinc practice

Current zinc practice includes roll cap roofing, the rolls having a tapered square profile, standing seam roofing and cladding with capped rolls or standing seams. The publication *Zinc in Building — Data Sheets* issued by the Zinc Development Association (see References) provides technical information for the laying and detailing of these roofs.

7.2 PROPERTIES AND PERFORMANCE OF ZINC

Zinc is a non-ferrous metal with an initially bright surface finish. During the first three to six months' exposure to the atmosphere, a dark matt grey patina mainly of zinc carbonate develops, providing an adherent and protective surface. The under surfaces of zinc sheet roofing can be subject to 'white rust', which develops from condensation. Contact between zinc and most metals is safe. However, copper and copper-rich alloys will cause electrochemical corrosion of zinc whether by direct contact or by water which has run off and contains copper in solution. Contact between zinc and iron is also best avoided.

Zinc is prone to attack by the sulphur products of air pollution, by acids and strong alkalis. Red cedar, oak and sweet chestnut can have significant corrosive effects on zinc by either direct contact or water draining from them. This effect is most severe in the first five to ten years while the corrosive acids are being leached out. Problems with timber may also occur where this is in damp conditions and where preservatives of the copper/chromium/arsenate types are used. An impervious layer may be required to isolate the zinc from the timber. Cement-based mortars do not cause problems, but walling materials which contain soluble salts such as chlorides and sulphates may do so. Zinc is resistant to rural and marine atmospheres but will corrode slowly in a high-sulphur industrial environment.

Zinc sheeting is substantially thinner than lead sheeting and, as life is relative to thickness, its effective life is shorter. However, it is still very significant and zinc of 0.8 mm thickness can be expected to give a maintenance-free life of about 40 years in average urban conditions, longer in rural areas and up to 60 years on steeply pitched roofs (reference 4, p.13). The life of a sheet of traditional ordinary commercial zinc with its variation in thickness could only be equal to that of the thinnest part. (Modern zinc is very uniform in thickness (see section 7.3).

Zinc is less malleable than lead and consequently more forming work is done off the roof. Before sheet metal forming machines were developed sheets were dressed on site on a special bench.

The general principles of good zinc work are as follows:

- The covering must withstand wind forces, particularly suction.
- The fixing must not cause distortion through relative movement of the metal and substructure.
- At the same time, the joints which allow for limited movement must remain weathertight.

Assuming proper detailing and workmanship the durability of zinc roofs is independent of the system adopted for laying. Zinc as a material and a sheet roof covering has been incorrectly blamed for many failures whose real cause has been poor original workmanship.

7.3 FORMS OF ZINC

Until very recently the most common form of zinc sheet was Ordinary Commercial Zinc. Today zinc alloy is the prevalent form. A flashing grade zinc alloy is also available. Zinc sheet is therefore readily available for repair or replacement work.

Ordinary Commercial — traditional zinc sheet

Traditionally zinc was cast in ingots and had a coarse crystalline structure. When the ingots were rolled into sheets, working properties and ductility were improved. Rolling caused the zinc crystals to be drawn out and elongated, leaving the materials with a slight weakness when folded sharply across the grain. As long as care was taken during folding no problem arose.

Ordinary Commerical Zinc was 98.5 per cent purity, as smelters could not achieve greater purity. It was regulated by BS 849:1939, which has been superseded. Zinc sheets were pack rolled and therefore varied considerably in gauge.

Zinc alloy — modern zinc sheet

New zinc sheet now available is zinc alloy produced from electrolytically refined zinc of 99.995 per cent purity, alloyed with titanium (for higher strength) and copper (for improved ductility). The final quality is considered superior to traditonal zinc. Modern zinc is cast continuously and rolled at a single thickness. It has a smaller, more even grain structure than traditional zinc, making cracking due to folding most unlikely. Zinc alloy also has higher tensile strength, greater resistance to creep and less thermal expansion. It has a far more consistent thickness and higher quality surface finish. Modern zinc meets the quality requirements of BS 6561:1985, which supersedes the now withdrawn BS 849:1939.

Modern zinc flashing

Flashing grade zinc is alloyed with a very small percentage of lead and rolled to a soft temper. It is available in one thickness only and is covered by an Agrément certificate.

7.4   REPAIR AND MAINTENANCE OF ZINC ROOFING

While the repair of zinc roofing is perhaps not as practicable as with lead, it is certainly possible and should always be considered as an option.

A sheet which is generally sound, and damaged in a limited area, can have a zinc patch soldered on. Some forms of physical damage may be suited to a solder repair alone. Because of the fire hazard provided by the inodorous felt and timber backing of zinc roofing and the transfer of heat problem, wherever possible the sheet should be disengaged, repaired and replaced. Overlapping the patch is a further and at times alternative precaution.

An antimony-free soft solder to BS 219 of either grade I (50 per cent tin plus 50 per cent lead) or grade K (60 per cent tin plus 40 per cent lead) should be used. A flux must be used, e.g. zinc chloride solution (killed spirits) or a mixture of 5 parts zinc chloride solution to 1 part ammonium chloride (sal ammoniac) or a proprietary non-acid liquid flux. Flux residues should be washed off after completion of the joint with warm clean water. Paste fluxes should not be used and soldering should be done with a soldering iron, preferably of the continuously heated type.

It is relatively easy to replace a zinc sheet which has been fixed with the roll cap system. The cap can be removed with normally no more damage than the loss of one clip. The sheet fixings can then be easily seen and removed. Replacement sheets should be of the same size and thickness of the original. If an area of roof needs to be replaced it is most important to repeat the sheet sizes and to resist the temptation to replace a bay in one long strip. Joint types and roll cap profiles should also be repeated.

It is difficult to remove any roof sheet material fixed by the standing seam system without causing damage. The seams are particularly difficult to unfold, especially at the upstands.

Nails and screws for fixing work should be galvanized steel to the British Standard.

The maintenance inspection of a roll cap zinc roof should allow for the removal of the caps. It will not be possible to see the deterioration within a standing seam.

The repair of a zinc roof will likely include work on the boarded structure and the inodorous felt underlay.

Zinc roofing was intended to be left unpainted and it is best that it remains so. Thick coatings of heavy bodied paint can block anticapillary spaces in joints and thereby encourage water entry. Should it be necessary to isolate zinc roofing, e.g. from water run-off from a lichen-covered roof, a good quality exterior paint should be applied in a thin coating. Zinc reacts with most oil-based paints, forming soluble salts under the film, which leads to embrittlement of the paint or loss of adhesion. Zinc primers should be used first to prevent this effect.

The best method of removing dirt from zinc is a bristle brush and water. Light abrasion with stainless steel wire will probably require that a whole area be done. A solution containing 0.5 per cent sulphuric acid and 5 per cent sodium or potassium dichromate, left on for 30 seconds and washed off with water, may also be effective and although it is not damaging it may blacken the surface.

If other materials adjacent to the zinc are being cleaned, the zinc must be protected, particularly if the cleaning agents are strongly acidic or alkaline.

REFERENCES AND FURTHER INFORMATION

References

1 British Standards Institution, BS CP 143, *Sheet Roof and Wall Coverings, Part 5: Zinc.*
2 British Standards Institution, BS 849: 1939*Plain Zinc Sheet Roofing* (superseded by BS 6561:1985).
3 British Standards Institution, BS 6561: 1985 — *Plain Zinc Sheet Roofing.*
4 Gayle, Margot, Look, David W and Waite, John G, *Metals in America's Historic Buildings — Uses and Preservation Treatments*, US Department of the Interior, US Government Printing Office, 1980.
5 Institute of Plumbing, *Sheet Roofing Data Book and Design Guide*, Technical Committee, Oxford District Council and the Institute of Plumbers, Oxford, 1978.
6 Institute of Plumbing, Oxford District Council, Symposium, April 1978, Paper No 2 by George H. Ridge of the Zinc Development Association.
7 Ridge, George H., 'History of Zinc Roofing: Development Through 150 Years — Impetus from the Railway Growth.' *Roofing and Building Insulation*, Issue 3, March 1958, pp 2–3.
8 Ridge, George H., 'History of Zinc Roofing: Nineteenth Century Developments — Its Decorative Value.' *Roofing and Building Insulation*, Issue 4, 1958, pp 10–11.
9 Ridge, George H., 'History of Zinc Roofing: 1920 to the Present Day.' *Roofing and Building Insulation*, Issue 5, 1958, pp 10–11.
10 Zinc Development Association, *Zinc in Building*: Data Sheet 1 'Roll Cap Roofing'; Data Sheet 2 'Standing Seam Roofing', ZDA, August 1971.
11 Zinc Development Association, *Metizinc*, ZDA, January 1986.

Further information

Zinc Development Association
34 Berkeley Square
London W1X 6AJ
Tel: (01) 499 6636

8 ZINC SCULPTURE AND ITS CONSERVATION

8.1 SOURCES OF ZINC SCULPTURE

Zinc sculpture and other decorative elements were cast in Germany as early as 1832. At the 1851 Crystal Palace Exhibition in London the advantages of zinc for sculpture were praised highly. By this time zinc sculptures were being produced at foundries in Liège, Paris, Berlin as well as the United States. Both cast zinc and zinc sheet were being used to form ornamental pieces including fountains. A particularly fine collection of seventeen sculptures originating from foundries in Paris and Berlin and dating from about 1870 can be found in the formal gardens of Osborne House on the Isle of Wight. These were originally treated with a bronze finish. By 1900 the unsuitability of zinc for outdoor sculpture had become known and it no longer enjoyed popularity as a sculptural/ornamental material (reference 2, page 151).

Today a legacy of architectural and sculptural zinc remains which needs to be looked after correctly. Sculptural zinc must remain the domain of the metals conservator, as a wide range of specialist analytical and craft skills are needed for appropriate measures to be taken.

Problems which may be encountered include corrosion pitting to a minor or a distracting extent, defective foundry joints, disfiguring repairs, corroded wrought iron armatures, corrosion due to a concrete or other type of core, fractures in zinc which is unable to support itself, and casting flaws such as areas of porosity or dross inclusion.

8.2 TREATMENT OF DETERIORATED ZINC SCULPTURE

Once a zinc sculpture has been inspected and its condition and causes of deterioration have been assessed, it should be removed to a workshop so the appropriate repairs can be undertaken properly. Most of the damaging and inappropriate works which have been undertaken on zinc sculptures have been *in situ* repairs.

Assessment of a zinc sculpture should involve analysis of the original metals and

The formal gardens of Osborne House on the Isle of Wight contain seventeen zinc sculptures produced in Germany and France. They were originally bronzed by electroplating to match the other bronze sculptures in the garden.

127

joint fillers, repair materials such as solder, and the products of corrosion. The analysis results must be interpreted correctly within the context of the sculpture.

In the workshop the sculpture will need to be dismantled along any unsound foundry joints. These and any fractures will need to be cleaned. Any filling or core material and any rusting iron armature or chaplets should be removed and the internal surfaces cleaned of corrosion products. Appropriate cleaning methods for the internal and external surfaces of zinc will include neutral pH soap in warm water applied with bristle brushes, non-caustic degreasing agents and organic solvents to low pressure abrasive blasting, for which the operating pressure, blast media (glassbeads, nut shells or other), nozzle type and size need to be very carefully selected. The selection of the best method or methods will need to be made after a close inspection of the surface and determination of the nature of the dirt and the products of corrosion.

Where structural strength is required repairs can be made with a welding rod comprised of 93 per cent zinc, 4 per cent aluminium, 3 per cent copper using a clean oxygen/propane flame not to exceed 800°F (427°C) and a hydrochloric acid flux (reference 1). The traditional method of repair was to use antimony-free solder of 50 per cent lead:50 per cent tin or 40 per cent lead:60 per cent tin, also with a hydrochloric acid flux. As this flux can dissolve zinc all excess needs to be rinsed immediately after soldering. The corrosive nature of the soldering flux has led to the use of epoxy resin adhesives to provide structural joins in zinc bound for external display. The use of epoxy resin also stems from a desire to avoid the use of heat more than is absolutely necessary. The exterior of joints repaired in this way are usually filled with epoxy resin-based filler, the surface of which is shaped to conform with the contours of the adjacent castings.

During the reassembly of the zinc sculpture distorted areas may need to be re-shaped. The sculpture may require an internal armature. If this is to be made in stainless steel it will need to be isolated from the zinc to avoid galvanic coupling.

Where zinc is badly deteriorated or missing, a section may need to be replaced. Models and casts need to be prepared so a new piece can be cast in zinc and then welded, soldered or glued back on. It may be appropriate at times for a section to be modelled *in situ* in a putty of zinc dust and epoxy resin.

Corrosion pits and other blemishes will need to be filled to prevent further water collecting and increasing in corrosive effect as it evaporates. The pits need to be filled individually to achieve a surface which appears slightly uneven and weathered. Puddled zinc (reference 1) will give the best colour match. Zinc dust with epoxy resin has also been used, but it is not as easy to apply or remove as zinc dust and polyester resin. This work can be very time-consuming. If it is necessary to use filler extensively the visual implications of this will need to be recognized.

Zinc castings in outdoor situations benefit greatly from a coating which isolates them from contact with moisture. A coating of wax or lanoline would need to be renewed annually. The merits of wax and lacquer coatings on exterior sculpture are discussed in Chapter 3.

REFERENCES

1 Gayle, Margot, Look, David W and Waite, John G, *Metals in America's Historic Buildings — Uses and Preservation Treatments*, US Department of the Interior, US Government Printing Office, 1980.

2 Naylor Conservation, *Report No 191 — Conservation Report on the Zinc Statues at Osborne House, Isle of Wight*, Report for RTAS, English Heritage, unpublished, Survey date: February 1987, Naylor Conservation, Telford, Shropshire.

3 Nosek, Elzbieta Maria, 'Conservation of Outdoor Zinc Monuments', ICCROM Conference, *Conservation of Metal Statuary and Architectural Decoration in Open-Air Exposure*, Paris, 6–8 October 1986.

4 Weil, Phoebe Dent, 'Problems in the Conservation of Zinc Sculpture in Outdoor Exposure', ICCROM Conference, *Conservation of Metal Statuary and Architectural Decoration in Open-Air Exposure*, Paris, 6–8 October 1986.

See also the Technical Bibliography, Volume 5.